The Hawks

THE HAWKS

A SHORT HISTORY OF THE 14TH/20TH KING'S HUSSARS

Bryan Perrett

with a foreword by
Major-General J. M. Palmer
Colonel, 14th/20th King's Hussars

PICTON PUBLISHING (CHIPPENHAM) LTD

1984

First published by Picton Publishing (Chippenham) Ltd 1984
ISBN 0902633 94 5

Photoset in AM International Plantin
Text paper supplied by Howard Smith Papers, Bristol
Printed in Great Britain by Picton Print
Citadel Works, Bath Road, Chippenham, Wilts
PP43473

Foreword
by Major-General J. M. Palmer, Colonel 14th/20th King's Hussars

I COMMEND THIS BOOK for a number of reasons both to past and present members of the Regiment and to the general reader. The author has from the outset recognised that the regimental spirit, which is so much a feature of the British Army and so much envied by other armies, has been developed to a remarkable degree by the 14th/20th King's Hussars. He has also recognised that a regiment's reputation is derived from the sum total of the individual effort of its members and that, whatever the odds or the difficulties with which they have been confronted, the Hawks have loyally tackled every task they have been set and seen it through to the end.

Rather that presenting a mere recital of events, Bryan Perrett has provided a readable and lively narrative spanning the 270 years of the Regiment's history, setting its battles in their historical context and describing in detail the manner of its evolution from the days of scarlet-coated Dragoons to the professional high-technology armoured regiment of today.

It is a story of a family not an institution. Like all good families, it is intensely proud of its forebears and their exploits and is determined to emulate them in their pursuit of individual excellence in their profession, whilst at the same time enjoying life to the full. The lesson of this book is that this and every future generation of British soldiers will succeed, providing that the regimental system is maintained, however untidy an organisation it may seem to those who do not understand it.

Contents

THE 14TH/20TH KING'S HUSSARS

The 14th King's Hussars

Orlando Norrie's painting of the 14th King's Hussars in line, advancing in review order.

Brigadier General James Dormer, who raised the Regiment in 1715.

The Jacobite Wars

THE REGIMENT came into existence on 22nd July 1715 and is recorded as having been raised in South Britain. This rather curious term has been interpreted as referring to the southern counties, from whence indeed some of the men may have come, but there is good reason to believe that the majority of recruits were drawn from a much wider area, for at the time the United Kingdom was divided into two Military Districts, North and South Britain, each with their own Secretary of State, the former including Scotland; a further reference to this period is contained in one of the titles belonging to the Royal Scots Greys, that of the 2nd North British Dragoons.

The men joining the Regiment included a high proportion of veterans who had seen active service in the long War of Spanish Succession, which had ended as recently as 1713. That they knew their business is beyond doubt, for when the Regiment was committed to action, only four months later, they acquitted themselves in a manner which would have been impossible for raw, partially trained troops.

The officer responsible for raising the Regiment was Brigadier-General James Dormer, who was also its first Colonel. The Regiment was officially listed as the 14th Regiment of Dragoons but, as was common during the period, was more often referred to by its Colonel's name. The uniform worn by Dormer's Dragoons consisted of a scarlet coat with yellow facings, tricorn hat and knee-boots. The dragoon was a soldier who was intended to combine the mobility of cavalry with the firepower of infantry, riding into action but fighting on foot, and to this end the trooper was armed with a musket and bayonet as well as a sword which could be used in either mounted or dismounted action.

Dormer's, together with several other regiments, had been raised to counter the ever-present threat posed by the House of Stuart, which had attempted a coup in 1708 and which, it was feared, would do so again very shortly. In September 1715, these fears became a reality when the Earl of Mar initiated a rebellion in Scotland in the name of James, the Old Pretender to the British Crown. There had, of course, always been support for the Stuart cause in Scotland, but there were plenty in England, too, who regularly drank the health of the 'King over the water'. The rising spread to Northumberland, where its most notable supporters were the Earl of Derwentwater and a rascally Member of Parliament named Thomas Forster, the latter's splendid opinion of himself as a soldier being equalled only by his total lack of ability. Somehow, Forster convinced his fellow Jacobites that he was the man to command the rebellion in England. His plan of campaign was to advance southwards through Lancashire, where he hoped support would be forthcoming, then capture the port of Liverpool and join forces with an army of Welsh Jacobites. As an essay in pure speculation this is hard to equal; the real danger for the Government was that the slightest success could provoke widespread unrest and encourage those who were wavering on the brink of joining the rebellion. By 9th November, Forster, who had been reinforced by Mar with several hundred men of Clan Mackintosh, had reached Preston and here he was joined by several Jacobite landowners, dragging their unwilling tenants with them.

Dormer's Dragoons were then at Manchester and were evidently still recruiting, since the Regiment's strength is given as only 231, or half the regular establishment. The Regiment formed part of a force commanded by Major-General Wills, which also included the 9th, 11th and 13th Dragoons, Pitt's Horse and three infantry regiments. Wills was a thoroughly professional officer who understood exactly how mounted troops should be handled and were he alive today he would probably do very well in command of an armoured battlegroup. He was determined to nip the English rising in the bud and set out at once for Preston, picking up Stanhope's Dragoons and a Militia battalion in Wigan. By three o'clock on 12th November his advance guard was approaching the bridge which crossed the Ribble to the south of Preston. This bottleneck, and the lanes leading to it, had cost Cromwell dear in 1648 and was now held by some of the best troops in the Jacobite army. In the normal course of events one might have expected this group to cover the deployment of the rebels along the north bank of the river, but Forster had completely lost his head on sighting the first dragoons. Mesmerised, he thought only of defence and had a series of barricades constructed in the town's principal streets; then, inexplicably, he withdrew his bridge guard.

Wills' regiments were soon streaming across the Ribble and very quickly had the rebels boxed in, blocking the roads leading east and north. This left Forster only one avenue of retreat, that to the west, which there were not enough troops to close; it was, however, an unattractive alternative which he chose to ignore. Wills, meanwhile, had decided to eliminate the rebels, rather than simply contain them, and launched concentric attacks into the town along the Longridge and Lancaster roads.

Dormer's Dragoons, together with the 9th and Stanhope's Dragoons, took part in the latter, thus going into action for the first time within sight of the area now occupied by the Regiment's Home Headquarters. Dismounted action was necessary and the horse-holders were told off. With tremendous élan the troopers stormed the first barricade in the teeth of concentrated musketry fire. Beyond lay a second and stronger barricade which was also attacked but which was stoutly defended. Had one or two guns been available this could probably have been knocked to pieces very quickly; unfortunately, they were not and, after a second attack had been held, Brigadier-General Dormer, commanding this sector of Wills' assault, pulled his men back behind the first barricade. Here they began to suffer casualties from snipers in the houses lining the street. Wills, not to be deterred, had the buildings set on fire.

By nightfall, both attacks had been halted and the battle had apparently subsided into stalemate. This was deceptive, for inside Preston the rebels' morale had been critically affected and Forster's life was in serious danger from his own followers. When, the following morning, Wills received just sufficient reinforcement to enable him to seal the western exit from the town, the Jacobites' resolve collapsed and they asked for terms, which were granted. On the 14th, Wills marched in to receive their surrender, taking 1,468 prisoners, many of whom were escorted by Dormer's to Lancaster Gaol.

It had been a brilliant action which completely stamped out the rebellion in England. The Jacobites, fighting behind defences, had seventeen men killed and twenty-five wounded; in contrast, Wills' troops sustained approximately 200 casualties, in itself a testimony to the vigour with which they pressed their attack. The Regiment's losses were remarkably light in the circumstances, amounting to three men and sixteen horses killed, and five men wounded. Among the latter was Brigadier-General Dormer himself, the first of a long line of Commanding Officers who were to be killed or wounded while leading the

The rebels lay down their arms at Preston under the watchful eyes of Dormer's Dragoons.

Harris Museum and Art Gallery, Preston

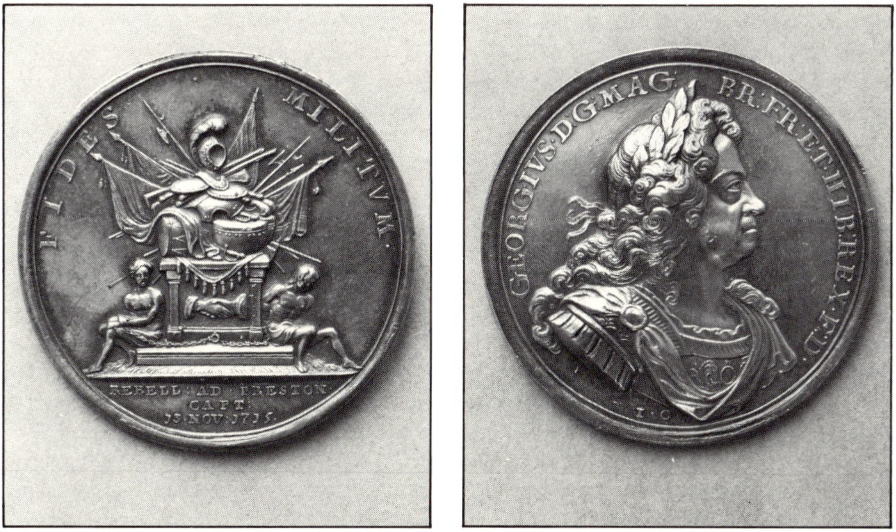

The first decoration to be awarded to members of the Regiment. Obverse and reverse of the rare commemorative medal struck for Preston Fight.

Harris Museum and Art Gallery, Preston

Regiment into action. This high standard of leadership has been maintained ever since and its essence is captured by one simple phrase – Follow Me!

While the rebels were requesting terms at Preston their hopes in Scotland had been dashed by the battle of Sheriffmuir. Here the Royal army, commanded by the Duke of Argyll, had fought Mar's in a remarkable encounter in which the right wings of both armies had simultaneously been victorious, the fact concealed from each other by a crestline. Technically, neither side was victorious, but it was Argyll's troops who remained on the field while Mar's, disheartened, began to disperse to their homes. Sheriffmuir, although of no direct interest to the Regiment, was to prove a baleful influence on its history a generation later.

The Jacobites were to attempt another insurrection in 1719, but this was quickly put down and the Regiment was not involved. It had, in fact, been posted to Ireland in 1717, and was to remain there for the next twenty-five years. During this period, the dismounted role of dragoons was largely abandoned throughout the Army, despite the object lesson of Preston. This was a matter of great concern to some, who saw the role as a necessary one. In 1728, Major-General Henry Hawley published a pamphlet on the subject, entitled *A Scheme for Reviving a Regiment of Original Dragoons for the Use for which they were First Intended, Since the Modern Dragoons are become Better Horse than ever was in England Before.* Hawley, a former dragoon, was an experienced cavalryman and, in his day, a very competent officer. He was also quick-tembered, foul-mouthed and a man who refused to alter whatever opinions he had formed, regardless of the facts. He was to give the Regiment little cause to remember him kindly.

In 1742 the Regiment, now known as Hamilton's Dragoons, returned to England and three years later moved to Scotland on the outbreak of the 'Forty-Five Rebellion'. This, the most dangerous of all the Jacobite insurrections, admittedly possesses the sentimental attraction of a lost cause, but has been romanticised beyond recognition. For the Army in general it was a black period in which it suffered a number of apparently inexplicable defeats. For Hamilton's, it was a particularly unhappy experience which has led to adverse comment ever since; some of this is inaccurate, and much of it is unnecessary. One pertinent fact, frequently forgotten, is that the majority of regiments which won the decisive battle of Culloden had also been worsted by the Young Pretender's Highlanders at one time or another.

Rather than dwelling on whatever deficiencies the Royal army might have had, therefore, a more positive approach is to examine the assets which its opponents possessed. Feudal levy though Prince Charles' army might have been, its Clan regiments were disciplined, fiercely motivated and intensely loyal to their Chiefs. Most Highlanders were armed with a musket, a broadsword and dirk, and carried a small shield on the left arm. Some, too, were armed with the fearsome two-handed Lochaber axe, which could cleave its way into a horse's neck without difficulty. Again, their methods were quite at variance with the formalised tactics of mid-eighteenth century warfare. First, with the aid of their pipers, they would rouse themselves to a state of berserk fury; then they would advance to within 100 yards of the enemy line, fire a volley and charge home at a run, screaming their Gaelic war cries. An attack by Highland infantry was – and remains – almost impossible to stop, as some of the best soldiers in the world have found to their cost. Nor, unless they had been routed, had they any fear of cavalry. Whereas other infantry might form square when charged, presenting the horsemen with a rectangular hedge of bayonets, Highlanders were content to receive the attack in line, opening fire at the last possible moment. Even if ridden down they would use their dirks to stab upwards into the horses' bellies, or slash with their broadswords at the animals' hamstrings. In the light of this, the events of the Forty-Five become more intelligible.

The Regiment's first operational task was to join the 13th Dragoons, known as Gardiner's Regiment, which had been screening the Jacobite advance on Edinburgh, at Corstorphine. The two regiments then retired to Coltbridge, just west of the city, on 16th September, where it was decided to continue the withdrawal. A line of picquets was put out as a precaution while they formed up and moved off. At this phase of the campaign the rebel army did not possess any cavalry of its own, but a few of its members were mounted and these were acting as scouts for the main body. They reined in on sighting the picquets and began to pop off their pistols, apparently to little effect. The picquets, having seen their regiments safely away, broke contact and wheeled about, the rebel scouts moving forward to occupy their position, from which they could see the dragoons' distant column. The rearmost troops were moving at the canter, but there was nothing sinister in this, and anyone who has ever travelled in convoy will know that those at the back frequently have to travel much faster than those who are leading. The rebels also saw the picquets galloping along the column to rejoin their own troops and, being enthusiastic amateurs, optimistically assumed that the dragoons were in headlong flight. Somehow, this perfectly routine cavalry manoeuvre was translated by the Jacobites into a famous victory and dubbed the Coltbridge Canter. Significantly, one officer riding with the column was Brigadier-General Thomas Fowke, Second-in-Command to General Sir John Cope, the Commander-in-Chief in Scotland, and an honest man who scrupulously recorded anything that was amiss. When questioned regarding the enemy's allegations of panic, he was completely nonplussed and was emphatic that the regiments' withdrawal had been 'regularly performed'.

Bypassing Edinburgh, which had decided to surrender to Prince Charles, the dragoons joined General Cope's army near Dunbar. The Jacobites, however, as well as having formidable troops, also possessed some capable officers, and during the night of 20th/21st September the rebel army marched round Cope's position at Prestonpans. A misty dawn revealed it drawn up and ready for action only 200 yards from the nearest troops. Cope had barely time to change his front, with Gardiner's and some artillery on the right, the infantry in the centre and Hamilton's on the left, before the Highlanders launched their attack. It is a matter of record that the battle lasted less than ten minutes and ended with the virtual disintegration of the Royal army. Nonetheless, the Regiment's part in the catastrophe does bear detailed examination.

Horsemen subjected to a determined infantry attack are actually at a serious disadvantage; this the British Army was itself to demonstrate at Minden in 1759, when a mere handful of battalions drove the French cavalry off the field. In this instance, the Highlanders slashed at the horses' legs and faces and, in such circumstances, it is hardly to be wondered that troopers soon found their mounts ungovernable, or that troop and squadron formation was quickly lost. Again, the enemy's right extended some way beyond the Regiment's left and this swung round to compress the ranks still further. The Regiment was now under simultaneous attack from two sides, and was virtually surrounded when the infantry in the centre broke, carrying away the reserve squadron with them. This impression is confirmed by an independent civilian witness to the battle, Lord Drummore, who had positioned himself, at some personal risk, only 150 yards from the Regiment's left flank. Hamilton's, wrote Drummore, left the field '*not* in a body, but quite broke in *two's or*

The Battle of Prestonpans, 1745, showing the effect of the Highlanders' attack on stationary cavalry.

so'. In other words, small groups of men were cutting their way out of what had become an impossible situation. Even more significant is a return submitted by Hamilton's shortly after the battle showing the loss of eighty-seven horses. Some of these were undoubtedly riderless strays, but the remainder can only have been killed during the minutes of furious close-quarter fighting; the terrible effect of the Highlanders' weapons was commented upon by all who visited the field, which was covered with bodies, severed limbs and maimed, dying or dead horses. One incident of which the Regiment has reason to be proud is the tremendous fight put up by Major Richard Bowles and a small group of troopers, all of whom were eventually cut down. Bowles, his horse having been killed under him, con-

Mary Evans Picture Library

tinued to fight on foot; he had sustained eleven wounds and was on the point of being hacked to death when an officer of Clan Cameron intervened to save his life.

On the right flank of the battle, the experience of Gardiner's Dragoons had been very similar. Colonel Gardiner, who, as a young officer had fought with his Regiment at Preston, was himself killed fighting courageously. The impression given by many historians is that both dragoon regiments left the field in headlong flight. In fact they rallied at a village, by coincidence also called Preston, which was only 1,000 yards behind the original line of battle, and here the Highland advance halted. With the exception of 170 infantrymen, they were all that remained of Cope's army; 300 of his men had been killed and some 1,400 were prisoners, of whom 500 were wounded. The rebel loss amounted to about thirty killed and seventy wounded.

Cope had no alternative but to retire to Berwick, where the recriminations began to fly. The dragoons' sharpest critic was a Colonel Lascelles, an infantryman. To him Lt-Colonel Whitney of Gardiner's, himself wounded in the battle, returned a very dusty answer to the effect that Lascelles was hardly in a position to pass judgement, 'the Action ending at Half an Hour after 5 o'clock, you was seen at the Post House at Haddington At half an Hour after 6, Eight long scotch Miles from the Field of Battle!'

The Jacobites now commenced their abortive march into England, while the Regiment joined a fresh army which was assembling at Newcastle. Operational command of this was given to Lt-General Henry Hawley, whose distinguished career to date could not conceal the fact that his judgement had become suspect with the years, nor that he had developed the definite psychopathic tendencies which led to his nickname of Hangman. He was still, however, an officer of some energy and pushed north into Scotland to re-occupy Edinburgh. Prince Charles' army, meanwhile, had returned from England and was besieging Stirling Castle, which Hawley was determined to relieve. On 13th January 1746 the Regiment took part in a small action which successfully ejected a rebel force from Linlithgow. The advance continued as far as Falkirk where, during the late afternoon of the 17th, reports were received that the enemy were moving onto the moor above the town. Why Hawley chose to engage with so little daylight left and in conditions of driving sleet remains a mystery. He seems to have been concerned with possession of the high ground, but the simplest reconnaissance would have revealed that the Highland army was already deploying on the summit of the moor.

Led by its three dragoon regiments – Hamilton's, Ligonier's (late Gardiner's) and Cobham's – the Royal army marched out of Falkirk and up onto the moor. The dragoons formed a screen behind which the infantry formed their line and then took up position on the extreme left. Hawley had fought on the victorious right wing of Argyll's army at Sheriffmuir and had formed an opinion that Highlanders would always break when charged by dragoons. There was nothing tangible to support such a view, but now he launched his cavalry in a charge at the enemy's right. To be effective any cavalry charge has to be delivered at the maximum possible speed with the troopers riding knee to knee. On this occasion, the squadrons attacked uphill and barely succeeded in working up a canter, their dressing being disordered by the rough going. The rebels held their fire until the horsemen were only ten yards distant and their volley cut down most of the leading rank. Here and there, little knots of officers and men broke into their line but were quickly pulled off their mounts and killed. Before the regiments could re-form the Highlanders swept forward in a counter-attack which soon extended along the whole front. This carried all before it save for three infantry regiments on the extreme right of Hawley's line, whose position was protected by a ravine and who acted as rearguard while the army streamed back into Falkirk. Once again, the Jacobites had won a quick, easy victory, killing over 300 of Hawley's men in the process, at a cost of fifty killed and eighty wounded.

Hawley, after smashing his sword against Falkirk Market Cross in a paroxysm of rage, withdrew to Linlithgow where he proceeded to vent his spleen on his troops. Thirty-two infantrymen were shot for alleged cowardice, but even this privilege was denied to thirty-one members of the Regiment who rejoined after the battle. Hawley had long had his knife into Hamilton's and they were hanged for 'deserting to the rebels'. This was, of course, the most arrant nonsense, for if the charge was justified they would surely have remained with the rebel army. These unfortunate men were almost certainly troopers who had been unhorsed and then captured during the Highlanders' counter-attack, being subsequently released as the Jacobite army was not equipped to hold its prisoners. Hangman Hawley, deaf to argument, was thus guilty of murder.

On 30th January, the Duke of Cumberland arrived to assume command from Hawley and commenced the march which would finally crush the Jacobites' ambition forever. It is interesting to note that at the decisive action of Culloden the Highland army fought a defensive battle and that its most serious counter-attack was foiled by tactics which the Duke had devised to deal with its unconventional style of fighting. Neither Hamilton's nor Ligonier's Dragoons were present at Culloden, the most probable reason being that Prestonpans and Falkirk had reduced their strength in men and horses below the level of operational efficiency. During this period their survivors were employed on what today would be called a route denial role, patrolling the roads between the Forth and the Clyde; this effectively isolated the Highlands from the rest of the country and prevented news or support of any kind from reaching the Young Pretender.

So ended the least rewarding period in the Regiment's history. Even if its efforts had not been crowned with success, the reasons for this lay beyond its control. That it had done its duty to the best of its ability is beyond doubt; had it been otherwise its casualties would hardly have been among the highest in the Army.

Trooper and Trumpeter, 14th Dragoons, 1742.

Conversion to Light Dragoons and The War with Revolutionary France

IN 1747, THE REGIMENT went to Ireland again, where it remained for the next forty-eight years. For most of this period its strength was that of a strong cadre, but it was efficiently run and attracted favourable comment when reviewed by senior officers. One slight personal reservation, expressed in 1773, was that 'the men in general ride too long', i.e. with the leg straight; many ascribe the origins of the present seat to the nineteenth century, but apparently it was finding favour somewhat earlier. In Ireland, the Regiment followed the normal routine of changing its station frequently, the object being to prevent troops becoming too embedded in, and therefore sympathetic to, the local population, against whom they might be called upon to act in aid of the Civil Power, a far from infrequent occurence. Throughout this period an unexpected strand of continuity was provided by the Regimental Chaplain, Mr Peter Vatass, who received his appointment at a salary of £120 per annum in December 1745, and then set what may be both military and ecclesiastical records by remaining on leave for the next fifty-two years!

In 1776, the Regiment joined what had become an élite branch of the cavalry, the Light Dragoons, and in the same year the practice of identifying it by its Colonel's name was discontinued. A move towards light cavalry had begun during the 'Forty-Five', and in 1756 a Light Troop was added to the establishment of every dragoon regiment, its function being reconnaissance. These Light Troops were often detached from their regiments and brigaded together, performing so well that in 1759 it was decided to raise permanent Light Dragoon regiments and, subsequently, to convert some existing dragoon regiments to the role. The ideal light dragoon is described as being no shorter than 5 feet 5½ inches and no taller than 5 feet 9 inches, 'and by no means gummy'. His mount was smaller than that of the conventional dragoon, standing between 14.3 and 15.1 hands, and was of 'the nag or hunter kind'. His training was more exhaustive than that given to the average cavalry trooper. He was taught to fight mounted or dismounted, to fire from the saddle, to skirmish, to demolish barricades, to throw up defences and, above all, *to use his initiative.* He went about his business at top speed, letting neither hedge, ditch nor river delay him. On active service he carried a sabre, pistols, carbine and bayonet, a hedging bill and a spade. He was just about the most versatile soldier in the Army but he was, first and foremost, a cavalryman doing a cavalry job, and that meant reconnaissance, forming screens, raiding, pursuit and, if necessary, charging in close order.

In addition to the obvious changes mentioned above, the 14th Light Dragoons exchanged their cocked hats for a low helmet of Roman pattern worn, on parade, with a trailing crest of white horsehair. In 1784 the scarlet coat was abandoned in favour of a dark blue shell jacket with yellow facings, trimmed across the chest with white cord, worn with the smart fur-crested Tarleton helmet.

War with Revolutionary France broke out in 1792 and the following year a British force under the Duke of York commenced operations against the French in the Low Countries.

14th Light Dragoons, Trooper, 1776.

In 1794, the Regiment sent two troops to join this army, and these were present at the engagements of Boxtell, Tuyl, Geldermalsen, Bueren and Elst. This campaign rivals the Crimean War as being the worst managed in the Army's history yet despite being left without food, clothing, medical services and pay, as well as suffering an acute shortage of ammunition, the soldiers did what was asked of them. The survivors of the expedition withdrew to Osnabruck in Germany, where the remnants of the Fourteenth's troops were absorbed by the 8th Light Dragoons, sailing for England from Bremerlech in April 1795.

Notwithstanding their sufferings in the depths of a harsh Continental winter, there were many who would regard these men as lucky, for, shortly after they had returned, the Regiment's seven remaining troops embarked for Haiti. In those days, a posting to the West Indies was regarded as a death sentence, thousands dying each year from tropical diseases of which the best known was the virulent Yellow Jack. It was because these diseases were so fatal to Europeans that the island plantations were worked by slaves imported from Africa. Haiti had been a French possession and, naturally enough, the slaves had followed the lead of the Paris mob in proclaiming Liberty, Equality and Brotherhood. They had risen, slaughtering their masters, and it was feared that their rebellion would spread to neighbouring British-owned islands, whose produce was a vital element in the United Kingdom's economy. It was, therefore, decided to bring the situation under control.

The terrain of Haiti was close and hilly, enabling the insurgents to mount regular ambushes and then retire into the bush, where the cavalry could not follow. To this galling type of warfare was added an outbreak of Yellow Jack which quickly decimated every British regiment present. On 2nd June 1797, however, a considerable body of insurgents was discovered drawn up and ready for action outside the town of Le Mirebalais. The British cavalry, consisting of detachments from the 14th, 18th and 21st Light Dragoons, immediately turned the enemy's right flank and charged. The insurgents, who had probably never witnessed such an event, bolted as fast as their legs could carry them, abandoning their artillery, ammunition and mules. In this instance the pursuit took place over open going, but the troopers were hard pressed to catch the fugitives. Many, however, were driven into the River Artibonite, where they drowned. This successful action was concluded at a cost of only two men wounded.

Shortly after Le Mirebalais, the Regiment embarked for England, arriving in October and going into barracks at Chelmsford. Of over 400 men who had sailed for the West Indies two years earlier, only twenty-five returned. This does not mean that the remainder were dead, but if the comparable experience of other regiments is considered, it suggests that almost 300 had succumbed to Yellow Jack and that 100, who were considered to possess an immunity, were posted to other units remaining in the islands.

Nonetheless, the Regiment recovered its strength in a remarkably short space of time and by 1798 could muster 600 men. During that year it provided an escort for HRH the Princess Frederica, Princess Royal of Prussia, on her arrival in England. The Princess had married the Duke of York, now Commander-in-Chief of the Army, in 1791, and evidently formed a most favourable impression of the Regiment, for on 26th July its title became 14th (Duchess of York's Own) Regiment of Light Dragoons. As a mark of special favour, Royal authority was given for it to assume the Prussian Eagle as a badge, this being worn on the right side of the helmet. The colour of the uniform facings was simultaneously changed to orange, this being the colour of the Brandenburg livery; the guidon, which until now had been crimson, also became orange with the Prussian Eagle in the centre, the White Horse of Hannover in the 1st and 4th quarters and 'XIV LD' in the 2nd and 3rd.

For the next ten years the Regiment was employed on garrison duties in southern England, but in 1808 it received orders to embark for Lisbon and the war which was to establish its reputation.

The Peninsular War

WELLINGTON'S PENINSULAR ARMY, like the 8th or 14th Armies in the Second World War or, more recently, the Falklands Task Force, possessed a personality that was uniquely its own. It fought a professional war against professionals in which both sides observed a gentlemanly code of conduct, revealing little personal animosity between enemies, but rather a great deal of mutual respect. Cavalry vedettes conversed regularly and, if a major action was imminent, their officers would shake hands before retiring to their own lines; if they met again after the battle, it was by no means unknown for them to share a bottle of wine.

Soldiers of every rank and arm of service have left their personal records of this war and the impression they give is of simple, straightforward men, completely immersed in their surroundings. In this respect the Regiment is particularly fortunate to possess the reminiscences of one of its troop leaders, Captain Thomas Brotherton, a former Guards officer who had transferred to the Fourteenth in 1807, by which date he was already an experienced soldier. Brotherton had an eye for detail which made him a fine *raconteur*, and he was undoubtedly a good companion who could be relied upon to profer his canteen, which he filled with 'strong wine', at suitable moments, as well as providing entertainment with his fiddle around the campfire. He tended to seek action rather than await its coming and complained little when injured, but was somewhat intolerant of those who might not share these views.

'One of the officers under my immediate command, Cornet B——, was hit by a fragment of a shell in the posterior, and as he was rather a soft sort of fellow, I thought, at first, that he made too great a fuss about it, though he turned deadly pale. But he had good reason to complain, for the piece of shell had buried itself deep in his buttock, and caused his death.'

Poor Cornet B—— was not alone in being dubbed by Brotherton as 'rather a soft sort of fellow', yet events could prove the Captain wrong and when they did so he was mortified by his mistake and promptly confessed the fact. But for all his fire-eating ways, Brotherton himself possessed a gentle nature and detested needless killing, particularly if the issue of the day had already been decided, and once exchanged sharp words with an officer of the King's German Legion on the subject. It would, however, be wrong to assume that he was in any way unique. As we shall see, there were many men of real ability, high courage and strong character serving with the Regiment, and indeed they tended to provide the rule rather than the exception.

The Crossing of the Duoro and The Battle of Talavera, 1809

IN EARLY 1809, A FRENCH ARMY under Marshal Nicolas Soult had compelled the British force commanded by Lt-General Sir John Moore to evacuate Corunna and had then advanced into northern Portugal, capturing Oporto. It was essential that the French by ejected and Wellington marched north from Lisbon to effect this.

Soult had, in fact, over-reached his resources and had already decided to withdraw when the approach of the Allied army was reported. He had the bridges across the River Duoro destroyed and all the local boats moved to his own bank of the river. This latter task was not completed efficiently, for when the Allies reached the Duoro on 12th May they discovered three barges. Wellington immediately used these to push some infantry across and sieze possession of a seminary. The French rearguard launched a furious attack on the building while the rest of their army could be seen withdrawing in the direction of Vallonga.

The Regiment, commanded by Colonel Samuel Hawker and forming part of Brigadier-General Cotton's light cavalry brigade, had meanwhile been ordered to provide two squadrons for a move against the enemy's flank. These, under Major Felton Hervey, led the advance of a German brigade and two guns to a ferry several miles upstream, the plan being that once across they would fall on the retreating column. The crossing, made in boats, took longer than anticipated so that when, eventually, contact with the enemy was made, it was with their rearguard. For some reason the officer commanding the flanking movement declined to engage, so that it was left to the two Fourteenth squadrons, one led by Hervey and the other by Major the Hon Charles Butler, to attack alone and unsupported.

The attack was made in column up a narrow lane, the walls on either side of which had been lined by the French with sharpshooters. At the top of the lane lay more open country, but the gap itself was held by the serried ranks of the rearguard, their bayonets levelled. On the other hand, an attack by cavalry possesses a moral force at least equal to the physical, as does an attack by armour today. The French rearguard, seeing that its fire could not halt the squadrons, disintegrated at the last moment and an instant later the troopers were among them, hacking and slashing. The impetus of the attack can be guaged by the fact that both French general officers present were very roughly handled; one, Moy, was cut down while the other, Delaborde, was bowled over. It is not known how many of the enemy were killed and wounded but it is thought that about 300 surrendered. The Fourteenth sustained the loss of twelve killed, one missing and twenty-two wounded, the last including four officers. Major Hervey lost his right arm, but insisted on remaining with the Regiment.

Having disposed of Soult, Wellington now had to retrace his steps to deal with a second French army which was menacing the Portuguese frontier in the Tagus valley. This was nominally commanded by King Joseph Bonaparte – although the major decisions were made by Marshal Claude Victor – and numbered approximately 46,000 men. Wellington's Allied army, 27,000 strong, effected a junction with a Spanish force of similar size at Talavera de la Reyna, and was here attacked by the French on 28th July.

Lt-Colonel Samuel Hawker,
the Regiment's first
Commanding Officer
in the Peninsular War.

The Spaniards could always be relied upon to be totally unreliable, and while a few of their units fought well, most fled in a complete panic as soon as the French appeared. This placed Wellington at a serious numerical disadvantage, particularly in the area of artillery, since he could only oppose thirty light guns to the eighty heavier weapons deployed by Joseph. In spite of this, repeated assaults all along the line were beaten back. In the afternoon, however, a dangerous situation developed in the centre. The Guards and the King's German Legion, having thrown back the most recent attack, followed up their success too far and were counter-attacked and driven back in some disorder. The position was restored by two decisive actions. First, the 48th (Northamptonshire) Regiment wheeled smartly out of line and began to punish the flank of the French advance with disciplined musketry. Then Cotton's brigade, consisting of the 14th and 16th Light Dragoons, charged into the now wavering enemy ranks, putting them to rout. The crisis had passed and the following day the French withdrew from the field.

In the circumstances, this being one of the bloodiest battles of the war, the Regiment escaped with the remarkably light loss of only three killed, eleven wounded and one officer taken prisoner. Colonel Hawker, who was among the wounded, was awarded a gold medal. He returned home shortly after the battle and command passed to Lt-Colonel Neil Talbot, a popular officer whom Brotherton describes as 'a delightful fellow, but singular and eccentric, particularly in his dress'.

Meanwhile Wellington, unable to get any sense at all out of the Spanish Supreme Junta, conducted a skilful withdrawal back into Portugal, where his army went into winter quarters.

The Battles of 1810

NAPOLEON, ANGERED by his Marshals' defeats, poured troops into Spain and very quickly removed the armies of the Junta from the board. In May, his Army of Portugal, commanded by Marshal André Massena, laid siege to the frontier fortress of Cuidad Rodrigo. Wellington, unable to mount a formal campaign because of his inferior numbers, nonetheless sent up the Light Division, with Cotton's brigade attached, to harass the besiegers and cut up the strong foraging parties which they had sent out into the surrounding countryside.

On 11th July, General Craufurd, the Light Division's commander, provoked a sharp action near the village of Sexmiro, not knowing that Cuidad Rodrigo had fallen the previous day. Normally a sound, if choleric commander, Craufurd was in one of his volcanic rages and his judgement was clouded. Having despatched two of the Regiment's squadrons on a wild goose chase he ordered the third, commanded by Brotherton, to charge some French infantry who had already formed square. This was bad enough, as the squadron had no artillery support, but in addition the enemy were thoroughly experienced troops who belonged to the 61st Regiment, one of the best in the French service. As Brotherton relates:

'It was lying down, concealed in some high-standing corn, and only rose up when my squadron came within pistol-shot of it, and was beautifully steady. We charged it most gallantly, but they fired a deadly volley into us, and half my men fell killed or wounded. Colonel Talbot had put himself at the head of the squadron along with me. Poor fellow, he fell pierced by eight balls, literally on the enemy's bayonets. The moment the square had fired into and so sadly crippled us, it moved off to join its support close by, and we were so shattered as not to be able to follow. So steady and cool was this little square, that though my horse fell, with the wounded, within two yards of their ranks, not a man moved out to bayonet me.'

Talbot, a resolute individualist to the last, had gone into action wearing an elegant pair of nankeen pantaloons. Major Hervey took over the Regiment and remained in command throughout the war, being promoted Lt-Colonel on 4th June 1811.

Having captured Cuidad Rodrigo, Massena was free to advance into Portugal. Wellington fought a successful delaying action at Busaco on 27th September, during which the Regiment was held in reserve. He then decided to withdraw behind the Lines of Torres Vedras, the deep defensive zone which he had had constructed between the Tagus and the sea. During the retreat the Regiment formed part of the rearguard and was involved in a number of brisk clashes with Massena's cavalry.

Yet even amid the deadly serious business of war there was humour to be found, as Brotherton discovered when he was suddenly surrounded by a crowd of nuns who begged him to save them from the French. 'They were,' he commented, 'Neither young nor handsome, but old and sallow, from penance and vigils, no doubt; so I resolved on the expedient of placing these poor distracted creatures (22 in number) *en croupe* behind as many dragoons. They had uneasy seats, but clasped the dragoons tightly round their waists, and we brought them safe into the Lines of Torres Vedras, to their great joy and the great amusement of all those who saw my convoy. Lord Wellington heard of this adventure, and was much amused by it.'

As the French closed in, it soon became apparent to Massena that the Lines were virtually impregnable. Worse, the country for miles around had been deliberately stripped bare of food and forage while the long supply line, stretching all the way back to France, was at the mercy of Spanish *guerrilleros*. Reduced by casualties, starvation, sickness and desertion, the once proud Army of Portugal was forced to withdraw.

Officer, 14th (Duchess of York's Own) Light Dragoons, 1798.

Officer, 14th (Duchess of York's Own) Light Dragoons, 1814.

Fuentes d'Onor, 1811

DURING THE SPRING of 1811, Wellington pursued Massena as far as the frontier and then laid siege to the fortress of Almeida. The Regiment, now brigaded with the 1st Royal Dragoons under Major-General Slade, formed part of the army's advance guard and throughout March and early April were regularly engaged with the enemy. At Venta de la Serra on 8th March, Captain Babington's squadron, supported by Colonel Hervey and the rest of the Regiment, charged and routed four squadrons of the French 11th and 26th Dragoons, capturing fourteen men for the loss of two.

One feature of the French Army of the period was its remarkable resilience. Towards the end of April, Wellington was astonished to learn that Massena, substantially reinforced, was actually advancing again and intended to break his own hold on Almeida. The two armies met at Fuentes d'Onor on 4th May and it was soon apparent that the French possessed greatly superior numbers, their cavalry being four times as strong as Wellington's. The battle reached its climax the following day; Massena's determined attacks were contained with difficulty but eventually repulsed with serious loss, and Almeida surrendered a day or so later. It had, however, been a close run thing and Wellington commented that had Napoleon been present the result might well have been very different.

The Regiment was engaged in a number of fierce encounters throughout the battle. At one stage, half of I Troop Royal Horse Artillery, commanded by Captain Norman Ramsay, was swamped by French cavalry as it withdrew. Notwithstanding his desperate predicament, Ramsay was not a man to be intimidated and Napier, the historian of the Peninsular War, describes what followed:

'A great commotion was observed in their [i.e. the French] main body. Men and horses there closed with confusion and tumult towards one point, a thick dust arose, and loud cries, and the sparkling of blades, and the flashing of pistols, indicated some extraordinary occurrence. Suddenly the multitude became violently agitated, an English shout pealed high and clear, the mass was rent asunder, and Norman Ramsay burst forth at the head of his battery, his horses, breathing fire, stretched like greyhounds along the plain, the guns bounding behind them as things of no weight, and the mounted gunners followed in full career. Captain Brotherton of the 14th Dragoons seeing this, rode forth with a squadron and overturned the head of the pursuing troops, and General Stewart, joining in the charge, took the French General Lamotte, fighting hand to hand.'

Ramsay's gunners had earned themselves that rarest of artillery distinctions, an Honour Title, and their descendents will naturally swear to this day that they had solved their problem themselves. Brotherton, on the other hand, speaks of 'the charge I made to protect and rescue Captain Ramsay's guns'. It matters not at all, for every man has his own perspective and in this, one of the most spectacular incidents in the Army's history, there was credit and to spare for all concerned.

It was not, however, only British artillery which could handle itself well in action. A little later in the battle, a French Horse Artillery battery unlimbered and opened fire on Slade's brigade at a range of only 200 yards. The battery was charged frontally by Captain Knipe's

Lt-Colonel Sir Felton Hervey commanded the 14th Light Dragoons for most of the Peninsular War.

R. Caton Woodville's spirited painting of the Battle of Fuentes d'Onor, showing Ramsay's Troop, RHA, breaking through the French cavalry. At this point Brotherton's squadron of the 14th Light Dragoons launched a counter-charge which turned the enemy back, and the start of this can be seen on the far right, although it is shown a little too far from the scene of the action.

National Army Museum

squadron of the Fourteenth, which suffered severely when the artillerymen quickly switched from round- to grapeshot, but nonetheless rode through the guns. The squadron was brought out of action by Lieutenant John Townsend, later to command the Regiment, as Knipe had been mortally wounded. Knipe had always maintained that a frontal attack on artillery was less dangerous than the manuals suggested, and despite his experience he continued to argue the point with Brotherton even as he lay dying. Ironically, in this instance he had had no choice in the matter.

It was about the time of Knipe's charge that Colonel Hervey narrowly escaped what could well have been a fatal injury. The previous day he had tucked a bulky tome entitled *Quenado's Works* into his sabretache and now, as Brotherton relates, 'an eight-pound shot entering the sabretache, went through the horse, and just appeared on the other side of his body, without coming through the skin, and it was evident that the thick book prevented it from going through and taking off Colonel Hervey's leg. His leg, from the blow, immediately swelled to an immense size, but though the horse fell down dead, and in the fall again hurt him, he would not leave the field, but had himself placed under a tree, where he remained during the remainder of the battle.'

Fuentes d'Onor was one of the Regiment's hardest earned battle honours and also a fine illustration of the professionalism of Wellington's army. When forced to give ground the cavalry and infantry covered each other's movements by bounds; when the French cavalry were repulsed by the infantry squares, the British cavalry rode out to complete their discomfiture with a counter-charge. In one such incident a squadron each from the Fourteenth and the Royals captured twenty-five French hussars who had themselves just taken prisoner a number of the Guards' skirmishers.

In recognition of the Regiment's role in the battle, Hervey was presented with a gold medal. Losses amounted to one officer, four troopers and six horses killed; five officers, thirteen sergeants, twenty-eight troopers and twenty-three horses wounded; and three men and four horses missing. The proportion of casualties among officers and senior NCOs in itself demonstrates the style of leadership expected.

The battle cost Massena his job and he was replaced by Marshal Auguste Marmont. Wellington, however, was hampered by serious supply and administrative difficulties and the rest of the year passed in indecisive fighting. The cavalry of both sides remained active and frequent clashes took place in the wide areas which separated the armies. On one occasion Brotherton was compelled to make a hasty and trouserless exit from a comfortable billet when he awoke to find that French cavalry had arrived in the village and were banging on his door; luckily they were only passing through and he was able to return and restore his dignity!

These affairs were no less deadly for being small and reached a climax in September. The Regiment was now brigaded under Count Arentschild with the 16th Light Dragoons and the 1st Hussars King's German Legion and at Espeja was ordered to dislodge a French cavalry unit which was holding some rising ground. The enemy were Napoleon's crack Polish Lancers and they chose to meet the Regiment's charge in their traditional way, at the halt and with their lances levelled in a bristling hedge. The dragoons found no difficulty in separating the lances with their sabres and then broke into their opponents' ranks, cutting down sixty of them and putting the rest to flight. In another of this brisk series of actions, 'a French officer, while striking at Felton Hervey, perceived he had only one arm, and with a rapid change brought his sword down into a salute and passed on'.

Peninsular War medals awarded to 14th Light Dragoons
and now in the Regiment's possession.

Salamanca and Blasco Sancho, 1812

THE FIRST PART of Wellington's 1812 campaign was devoted to the capture of the frontier fortresses of Cuidad Rodrigo and Badajos. After these operations had been successfully completed he then advanced across Spain again, besieging Salamanca, while Marmont concentrated his forces for what he believed would be a battle resulting in the decisive defeat of the British and their Allies.

The Regiment was still brigaded with the 1st Hussars King's German Legion, although the brigade had been taken over by Major-General Baron Victor von Alten, and was heavily engaged during the manoeuvering which took place in the days preceding the battle. Brotherton, now a Major, was challenged to single combat by a young officer of the 14th Chasseurs à Cheval which, by coincidence also wore orange facings, and the skirmishers of both sides ceased firing while they watched the contest. Brotherton was much the older and more experienced hand, but the Frenchman was a canny fighter who kept the middle ground and made Brotherton's horse circle his own, aiming cuts at its head. While engaged in their swordplay, the combatants exchanged cheerful civilities until each simultaneously penetrated the other's guard, Brotherton receiving a cut on the bridle hand while thrusting into his enemy's stomach. The Frenchman returned to his own lines and Brotherton was genuinely upset to learn the following day that his wound had proved mortal.

A few days later Brotherton was himself run through in a furious melée with the enemy's dragoons, fortunately without damage to a vital organ. On the morning of the battle of Salamanca - 22nd July 1812 - he nonetheless reported for duty. Horrified by his appearance, Hervey threatened to have him arrested if he did not leave the field immediately. He obediently left the Regiment but promptly joined in someone else's battle, in which his horse was shot under him; after that, he admitted a little diffidently, 'I felt very unequal to further exertion'.

At Salamanca, Wellington played a waiting game, but one in which he manoeuvred as the situation demanded. Mistaking an adjustment of position on the British right flank for the start of a general retreat, Marmont launched an attack with his own left wing, which became seriously over-extended. Siezing the opportunity, Wellington counter-attacked with his 3rd Infantry Division and a brigade of Portuguese cavalry, while von Alten's brigade covered the extreme right of the line. For most of the way, the approach march of the counter-attack force lay in dead ground and, taken by surprise, the leading enemy division was thrown back after a brief but intense struggle. A French dragoon brigade intervened to cover its withdrawal and was counter-charged by the Fourteenth and the German Hussars and driven off the field. Having defeated Marmont's left, Wellington now proceded to roll up the rest of the French line. Marmont himself was wounded and what remained of the Army of Portugal was expertly withdrawn by his deputy, General Clausel.

It had, Wellington remarked, taken forty minutes to beat 40,000 Frenchmen; the Regiment, which had sustained the loss of only four killed and six wounded, had not returned from its pursuit of the enemy dragoons before the battle ended, and Hervey was presented with a further gold medal in recognition of its services.

Four days later a patrol, commanded by Corporal William Hanley, consisting of four troopers from the Fourteenth and four German hussars, captured three French dragoons near a hacienda at Blasco Sancho. Inside the hacienda were more of the enemy, making themselves comfortable for the night. There was only one entrance, which Hanley quickly secured, ordering his men to fire into the courtyard beyond. By means of a French-speaking hussar, he informed the dragoons' commander that the remainder of von Alten's brigade would be arriving shortly and that if he did not surrender at once the building's thatched roof would be set alight. The French, twenty-seven in number, gave up, emerging mounted one at a time to be stripped of their swords and firearms. Hanley formed them into fours, riding with crossed sturrups, with the patrol around them. They had hardly moved off when they met a cheerful Lieutenant-Colonel who imagined, understandably, that the respective British and French positions were reversed. 'Good-day, Englishman!' he called as he passed, cordially slapping Hanley on the shoulder; Hanley's response was to snatch the Colonel's sword from its scabbard and advise him that he, too, was a prisoner. Shortly after, the Colonel's orderly, with two baggage mules, was also snapped up, and the the little column made its way back to the British lines.

Wellington, delighted by the incident, gave Hervey a sum of money to be distributed among the patrol, who also benefited financially when the French horses were bought in for service with the brigade. Hanley was presented with an inscribed silver medal by the officers of the Regiment at a full dress parade.

In August, Wellington occupied Madrid and then went on to besiege Burgos, which proved too tough a nut to be cracked quickly. By October, however, the French had concentrated several armies against him and he was compelled to raise the siege and withdraw, for the last time, to the Portuguese frontier for the winter.

The Regiment again formed part of the army's rearguard and was in action throughout the retreat. In one incident Hervey, surrounded by the enemy's lancers, was saved by the intervention of his orderly, Sergeant Puss, who cut a way through for the Colonel, losing his own life in the process.

It is not clear quite when, but at some time during the 1812 campaign the Regiment was issued with a new uniform. The shell jacket was replaced by a coatee with wide turnback lapels of the facing colour, and the helmet gave way to a shako. In the event, the new uniform proved to be less serviceable than the old one.

Vittoria and the Invasion of France 1813–1814

THE EVENTS OF 1812 had provided a decisive turning point in the Napoleonic Wars. The Grande Armée had perished in the snows of the Russian winter and, one after another, Napoleon's allies had turned against him. Again, while Salamanca had convinced King Joseph Bonaparte that a military solution to his problems in Spain was no longer probable, the denial of further reinforcements following the catastrophe in Russia made his defeat inevitable. The British government, sensing at last an end of the long war with France, decided to provide Wellington with everything he needed to finish his task. Thus, the campaign of 1813 began with Wellington's army advancing from Portugal full of confidence while Joseph's, its morale understandably low, embarked on a slow withdrawal from Spain, encumbered by the booty of several years' occupation of a conquered land.

Wellington's advance was made rapidly and in three columns, with the Regiment forming part of the advance guard of the centre column. There were brushes with the French rearguard at Salamanca and at Burgos, where the fortifications were abandoned and blown up by the enemy. During the latter event, which took place on 12th June, Captain Milles' squadron charged and captured a gun.

The Allies caught up with Joseph's army and forced an engagement on 21st June. Wellington chose to attack on this occasion, throwing the French off their strongest positions and then turning their left flank to drive them in disorder from the field. The Regiment acted in support of the movement against the enemy left, but the terrain was so rugged and broken that there was no opportunity for mounted action. The crowded nature of the battlefield, coupled with the fact that the French abandoned all their artillery and baggage, inhibited any immediate exploitation, but the Regiment was soon involved in a pursuit which was continued as far as Pamplona.

Vittoria was one of the most rewarding battles ever fought by the British Army. The military booty alone amounted to 151 guns, 415 caissons containing 14,249 rounds of artillery ammunition, two million musket balls, 40,683 lb of gunpowder, fifty-six forage waggons and forty-four travelling forges. Among the abandoned waggons and coaches there were riches for the taking – jewellery, silver plate and pay chests filled with doubloons. However, while other units set to a'plundering, Hervey's discipline was tight and he insisted that the Fourteenth should not divert from the business in hand. The men would have been less than human if they had not felt that they were being deprived of their fair share, but Hervey, a shrewd student of human nature, was determined that they should have some reward for their years of effort. In those days the regimental farriers jogged along behind the squadrons in much the same way as do the fitters' vehicles today. They were regarded as sharp characters who could be relied upon to help themselves to whatever was going, and were known to conceal looted valuables in their churns, which were leather saddle buckets designed to hold their tools and medicines. Arriving at Pamplona, the Regiment halted and was off-saddling when Hervey had *Rouse and Assembly* sounded. During

National Army Museum

A party of the Regiment looting the French baggage after Vittoria. The soldier in the centre is shown holding Marshal Jourdain's baton. Ackermann's prints were noted for their attention to detail and the men are correctly shown wearing waterproof shako covers and leather inserts and trouser bottoms.

The Emperor, King Joseph's silver chamberpot captured at Vittoria and still used as a toasting cup on Regimental Guest Nights.

the subsequent parade, the farriers were called out and ordered to empty their churns in public, much to the delight of everyone. The resulting pile of plunder was auctioned and the proceeds distributed among the Regiment as a whole.

King Joseph had escaped from the battlefield in his coach which, together with its fittings, had been a present from his brother, the Emperor Napoleon. The coach was overtaken and abandoned in the pursuit and Joseph made off on a troop horse. His elegant silver travelling chamber-pot, together with its linen napkin, was removed by one of the Regiment's patrols, and in consequence the Fourteenth were soon known as The Emperor's Chambermaids by the rest of the Army, a distinctive nickname which is still in use. The chamber-pot itself, christened The Emperor, has accompanied the Regiment in all its travels and appears on special occasions in the officers' mess where, as an earlier historian commented, it 'is seldom seen except when filled to the brim with the choicest brands of champagne, and in this condition has passed through successive generations of the Regiment'.

Shortly after Vittoria, the Regiment was brigaded with its old comrades from the Jacobite Wars, the 13th Light Dragoons, first under the command of Major-General Long and later under Colonel Hussey Vivian. While the remnants of Joseph's army retreated through the Pyrenees, the Fourteenth continued to snap up prisoners, caissons and sorely needed supply convoys, arriving in France on 10th November.

Napoleon had meanwhile despatched Marshal Soult to restore order and it was soon clear that the French soldier fighting for France on French soil was a very different individual from the man who had fought for a lost cause in a country he detested. The Regiment was involved in some of the fiercest engagements of the entire war, and on one occasion its baggage was captured, but only after the overwhelmed escort had gone down fighting.

On 13th December, Major Brotherton was captured during an ill-advised venture. He was due for his brevet Lt-Colonelcy, which Vivian was prepared to recommend when a suitable occasion presented itself. The brigade commander suggested that he attempt to sieze a narrow bridge at Hasparren by *coup-de-main*. The affair went badly wrong and Brotherton suddenly found himself fighting a lone battle against the 2nd Hussards de Chamborants *and* the 13th Chasseurs à Cheval. He was wounded repeatedly – his life being saved by a leather cuirass which he had had made after his experience at Salamanca – trussed up and bundled off to the rear. His predicament worried him less than the loss of his magnificent Arab mare Fatima, one of the Regiment's most famous horses, and he offered to buy her back from his captors; the French, however, knew her worth and she was sent to the Imperial stud.

For a while Brotherton was quartered with a wealthy cutler in Tarbes, who treated him very badly. Wellington's continued advance meant that he had to be moved still further to the rear and before he left, the cutler, suddenly servile, asked him to write out a protection pass which he could show to British troops. Having several scores to settle with the man, who spoke no English, Brotherton obliged as follows:

'The bearer of this is one of the greatest brutes that ever breathed. He treated me like a dog during the time I was lying wounded in his house, and I strongly recommend him to the special notice of all British soldiers who, no doubt, will treat him as he deserves for his infamous conduct of their wounded comrade.'

It was the cutler's ill luck that the Regiment entered Tarbes first, and he received a first-rate hiding from its angry troopers.

The Battle of Orthez, during which the Regiment threatened the French lines of communication so seriously that Marshal Soult was forced to abandon a naturally strong position.

After a brief pause during January 1814, fighting began again the following month. The 13th and 14th Light Dragoons were now known as the Ragged Brigade, the months of hard campaigning, most of it in contact with the enemy, having reduced their uniforms to tatters. Nonetheless, at the battle of Orthez, fought on 27th February, they discovered a ford which enabled them to pass round the French left and menace a bridge which lay on Soult's line of communication. The bridge was well protected by artillery beyond the river and could not be taken, but the threat posed by the brigade forced Soult to abandon his positions and withdraw. A later generation would immediately recognise in this the tactics of the Indirect Approach, an essential ingredient of the *Blitzkrieg* technique.

Rain fell incessantly as the brigade, now part of General Fane's cavalry division, followed up the French withdrawal, delayed constantly by the enemy's demolitions. Toulouse, the final action of the war, was not fought until 10th April. In this, a largely infantry battle, the Regiment was not heavily engaged, but later escorted Wellington when he entered the city. Napoleon had, in fact, abdicated four days earlier, and hostilities formally ended on 18th April.

The Regiment embarked for England in May. It had gone to war 770-strong and returned with 561 officers and men; during its five and a half years in the field the depot had provided 445 reinforcements and its 'non-effectives', i.e. deaths from all causes and men too seriously ill or badly injured for further service, amounted to 654, which can be regarded as low given the nature of the war and the conditons in which it was fought. It had been superbly led throughout, possessed a tremendous *esprit de corps*, and had earned the

14th Light Dragoons Regimental Medal,
awarded during the Peninsula War.

Battle Honours Douro, Talavera, Fuentes d'Onor, Salamanca, Vittoria, Pyrenees, Orthes and Peninsula.

The Peninsular War Medal was not awarded until 1848, by which time many of the veterans had died or gone abroad or were simply unaware that they were eligible for it; whatever the reason, less than 200 of them received the medal. The clasps for which the Regiment was eligible were 'Talavera', 'Busaco', 'Fuentes d'Onor', 'Badajos', 'Salamanca', 'Vittoria', 'Pyrenees', 'Nivelle', 'Nive', 'Orthes' and 'Toulouse'. A goodly proportion of the recipients received all eleven clasps and a majority of the remainder more than half; one man, who had transferred from the Royal Wagon Train, was additionally awarded the 'Corunna' clasp; another is listed as having served under an *alias*, a not infrequent occurrence.

New Orleans and The Waterloo Despatch, 1815

GREAT BRITAIN had also been at war with the United States since 1812 and, as a direct result of the reputation which it had earned in the Peninsula, the Regiment was required to provide two squadrons for service in America. There was undoubtedly a need for cavalry properly employed, and the lack of it had compelled General Ross to mount a number of his artillerymen during the Chesapeake Bay campaign of 1814, one result of which was the burning of the White House and other public buildings in Washington. However, the Fourteenth's squadrons, less their horses, were bound for an attack on New Orleans and it remains a complete mystery as to what dismounted light dragoons were expected to accomplish in the Louisiana swamplands. The whole enterprise was typical of several contemporary amphibious operations mounted against the North and South American continents, in which the Admirals tended to outrank the Generals and set the objectives, with the result that the Army could find itself in difficulties that were not of its own making.

The Americans relied almost exclusively on militia regiments for their defence and, knowing that these would stand no chance at all against Peninsular veterans, placed them behind earthworks covering the route to New Orleans. The British advance was harrassed

His Grace The Duke of Northumberland

The village of Waterloo, the day after the battle. Wellington drafted his despatch in the large house on the left, and outside can be seen the chaise in which Major Percy began his epic journey to London; the captured Eagles can be seen flying from the windows of the vehicle.

by two American warships on the Mississippi and there was an entrenched battery across the river which enfiladed the approach to the enemy's main position. One of the warships was destroyed and the other retired upstream, but the battery was not cleared until after the assault had gone in on 8th January 1815. Needless casualties were incurred and the operation was abandoned, although the American commander, the future President Andrew Jackson, wisely made no attempt to interfere with the withdrawal. Of the Regiment's part in these events the commander of the troops serving with the fleet, Major-General Lambert, commented: 'The conduct of the two squadrons of the 14th Light Dragoons, latterly commanded by Lieutenant-Colonel Baker, previously by Major Milles, has been the admiration of everyone, by the cheerfulness with which they have performed all descriptions of service.'

The squadrons commenced their voyage home in March, the frustrations of the episode being aggravated by the discovery that it had taken place *after* a peace treaty had been concluded with the United States, a fact unknown to either side at the time.

With so much of its strength temporarily absent in America, the Regiment did not take part in the Waterloo campaign although two of its members, Colonel Hervey and Major the Hon H. Percy, were present at the battle, both serving on the Duke of Wellington's staff. Percy had been present at the ball given by the Duchess of Richmond and reported for duty in his ADC's dress uniform, having hastily pulled on his boots over his silk stockings. He was the only one of Wellington's Aides-de-Camp who was not killed or seriously wounded in the fighting, although he received a graze on the foot, and after the battle the Duke asked him to convey his official despatch to London, together with two of the captured French Eagles, which were to be presented to the Prince Regent. Percy stuffed the despatch into a velvet sachet, which had been given him for luck by his partner at the ball, snatched up the Eagles and set off for Ostend in a fast chaise. At Ostend he boarded the brig *Peruvian*, but when the wind dropped within sight of Dover he transferred to the captain's gig and had himself rowed ashore, doing his fair share at the oars. From Dover he set out for the capital by post chaise, driving hell-for-leather with three changes of team *en route*. At every town and village the post boys yelled the news as the chaise tore through. In London, Percy reported first to Lord Castlereagh and then traced the Prince Regent to a ball in St James's Square. Now, reeling with fatigue, his eyes red from a week without sleep and his uniform torn and bloodstained, he crossed the ballroom to drop the Eagles at the Prince's feet and hand over the sachet containing Wellington's despatch. His words, 'Victory, Sire – victory!' signalled the end of a war which had lasted more than twenty years. The Prince rewarded him with promotion and made him a Commander of the Order of the Bath. The despatch, the Eagles, Percy's uniform and the lady's sachet all still exist.

His Grace The Duke of Northumberland

His Grace The Duke of Northumberland

The ADC's uniform worn by Major Percy at Waterloo and while delivering the despatch.

The lady's sachet in which the Waterloo Despatch was carried.

The Duke of Wellington and those members of his Staff who fought in the Peninsular War, painted in 1850. Brotherton, by now a Major-General and Inspector General of Cavalry, is the third standing figure from the left.

The Marquess of Londonderry

The Royal Regiment

DURING THE IMMEDIATE post-war years there was a suggestion that the Regiment should be converted to lancers, a type of cavalry which was rapidly gaining popularity within the Army. No doubt the suggestion was made as an additional recognition of the Fourteenth's distinguished record in the Peninsula but the Regiment itself, mindful of the carnage it had inflicted on the wretched Poles at Espeja, was distinctly cool in its response and the idea was quietly dropped.

The Regiment remained at various home stations until 1841, being occasionally called upon to act in aid of the Civil Power. In 1820, for example, it found itself chasing smugglers on the Kent coast. This difficult and thankless task was made all the harder by the fact that the local communities depended on smuggling for their income, so that any form of intelligence gathering was virtually imposible; indeed, whatever information was received by the military was intentionally misleading. It was, too, an era of general unrest brought about by the social effects of the industrial and agricultural revolutions and by pressure to expand the franchise, and disturbances were commonplace. The most serious of these were the notorious Bristol riots of 1831, which resulted in widespread destruction of property and loss of life, aggravated by feeble handling of the situation on the higher level. One squadron of the Regiment was present and, once stronger personalities had assumed control, was permitted to handle the matter as it demanded, charging and dispersing the rioters in very short order.

Other duties in which the Regiment was regularly involved were the provision of guards and escorts for various members of the Royal Family, with whom it seems to have been extremely popular. In 1830, King William IV commanded that henceforth the Fourteenth should be known as the King's Regiment of Light Dragoons, according it a second badge of the Royal Cypher within the Garter. Within the Regiment, the old order changed slowly but inevitably. Colonel Sir Felton Hervey died in 1819; the sword presented by his officers to this outstanding leader and truly courageous man now hangs in a place of honour within the officers' mess. The following year Lt-Colonel Brotherton, who had returned from his brief period of captivity following the French surrender of 1814, left the Army on half-pay to attend to personal concerns. There were, on the other hand, plenty of old Peninsular hands left among the younger officers, sergeants and junior ranks, and it was these men who saw to it that when the time came for the Regiment to fight again it would do so with the same dash and efficiency which had prevailed against Napoleon's marshals.

During this period military tailoring passed through its most extravagent phase. The reasonably comfortable jacket of the later Peninsular years was replaced by an extremely tight coatee with a high standing collar, while the shako became a tall bell-shaped structure which even chin straps and cap lines could not always keep on. The orange facings became scarlet on the Fourteenth becoming a Royal Regiment, but changed to blue when the light cavalry went into scarlet uniforms between 1831 and 1840; thereafter, a final reversion to blue with scarlet facings was made.

In 1832, the old orange guidon was replaced by one in the traditional crimson of a Royal Regiment, with the Prussian Eagle in the second and third quarters, the White Horse of Hanover in the first and fourth and the Crown and Garter surrounding the legend 'XIV KING'S LD' in the centre. The new guidon, however, had a very short active life and in 1834 was, in common with those of all light dragoon, hussar and lancer regiments, withdrawn from service, the theory being that the dispersed nature of light cavalry duties was incompatible with the basic purpose of a standard. It would be 127 years before the next guidon would be trooped before the Regiment.

Whatever, and no matter how inconvenient, the uniform it was wearing, nor indeed whether the guidon was on parade or not, the Regiment presented a dazzlingly smart appearance at this period of its history, and in itself this tended to attract recruits. A Private Hall confesses in his journals - now in the National Army Museum - that his decision to join was greatly influenced by the prospect of wearing the uniform, so much so that he asked the recruiting sergeant to tell him more of the life he intended to lead. Needless to say, he was promised the jolliest of times, money to jingle in his pocket and a horse to ride like a gentleman. When he enlisted in December 1824, the reality was that the NCOs ruled with an iron hand, that the fine uniform he admired so much had to be bought with stoppages from his pay, and that much of a cavalryman's time was absorbed in heavy and frequently dirty work. Having come to terms with the fact that a soldier's life was hard, Hall settled down and remained with the Regiment for twenty-two years. His journals provide a startlingly detailed insight into life within a cavalry regiment both at home and in India.

Officers, 14th King's Light Dragoons, in un-dress and review order, 1835.

1798.

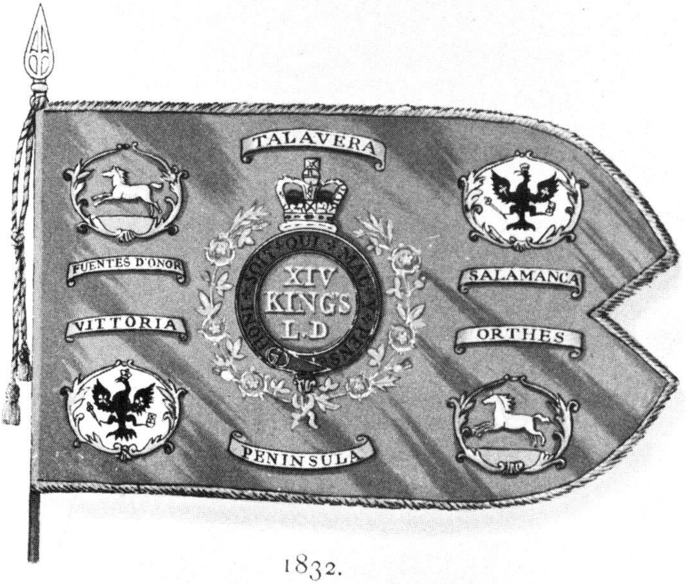

1832.

The Guidons of 1798 and 1832. The 1798 Guidon was in the regimental colour, orange; the 1832 Guidon being the King's Guidon – of which there was one per regiment – was crimson. Both were made of silk. In 1832 the King's crest – the Royal Cypher within the Garter – became for a time the principle badge in the place of the Prussian Eagle. Guidons were issued on a scale of one per two troops and until 1822 were carried by Cornets (2/Lts) and then by Troop Sergeant Majors. The use of Guidons was discountinued between 1834 and 1961. The battle honour 'DOURO' (1809) was not granted until 1837 and did not therefore appear on Guidons during that period.

John Company's India

THE REGIMENT LANDED at Bombay in September 1841 and marched to Kirkee, where it was joined by 150 men from the 4th Light Dragoons who had elected to transfer rather than return home with their regiment. One reason for this was given by Private George Tookey, who enlisted in the Fourteenth five years later and whose letters to his family have been preserved. There was never any shortage of inexpensive local labour willing to make the soldiers' lives a little more comfortable and, as Tookey put it, 'we live like gentlemen here'.

Conversely, the only concession to the intense heat was a white shako cover; in all other respects the troops went about dressed in the same uniform they wore in the United Kingdom. Death was an ever-present companion, appearing most frequently in the form of dysentry and cholera, so that regular drafts were required to keep units up to strength. Medicine was as yet a terrifyingly inexact science. For example, in 1846, Hall, now in his middle years, was incapacitated by symptoms which today would at once be recognised as those of a heart attack. The surgeon's remedy was to bleed him from both arms and then apply two dozen leeches to his chest. This reduced the pain but did little for his general condition and, shortly after, he was invalided out of the service. In other respects, when the Regiment was not engaged in operations or manoeuvres, boredom was the principal enemy.

India at this period was administered by the Honourable East India Company, whose essential interests were commercial. The Company rented British regiments from the Crown as a stiffening for its own Bengal, Bombay and Madras armies, of which the Bengal army was almost permanently engaged in operations of one sort or another. The bulk of its troops were locally raised regiments led by British officers who received their commissions from the Company, but there were also a number of intentionally named European cavalry, infantry and artillery units. Service in John Company's armies meant virtual exile and it may be wondered why it was so popular. For the officers, the answer was simple. In the Queen's Service, as they termed the British Regular Army, commission was by purchase and would remain so until 1870. In peacetime, promotion was also by purchase, although when a regiment was on active service this system was sensibly abandoned in favour of vacancy and merit. The difficulty was for a young man without means to get his foot on the first rung of the ladder, and the Company provided not only an answer to the problem, but also better rates of pay. Thus in 1855 a Cornet serving on one of the Company's cavalry regiments received £1.0.5d (£1.02) per day, while his equivalent in the United Kingdom received only 8/- (40p). There were, too, a number of ways in which one could make money in India, provided one was not too fastidious about one's methods. Most Company officers served for far longer than their counterparts in the Queen's Service, and since promotion was by strict seniority within regiments, it moved at a snail's pace. In 1855, Majors in the Bengal Light Cavalry were aged approximately 52, had held their ranks for 3½ years, and served an average of 30 years; their Regular Army equivalents were aged 37, had held their ranks for 3 years and served only 17 years. In the Company's service, Colonels tended to be in their sixties and Generals in their seventies, a situation bound to lead to disaster.

The men serving in the ranks of the Company's European regiments had no such incentives to lead them into lifelong exile, although again their rates of pay were better than could be obtained in the Queen's Service. Many enlisted to escape the consequences of their crimes, their debts or their wives, and there was always a sprinkling of adventurers hoping to make their fortunes by devious means. The majority of the men were British, but a number of foreigners were present as well. They were a hard lot but their very isolation produced an élitism similar to that of the French Foreign Legion, and they invariably fought well. By and large, however, the relationship between the Queen's and the Company's troops is adequately described as being distant.

The Regiment arrived in India while the First Afghan War was in progress, and, during this, British prestige took a sharp tumble when a complete column was wiped out in the Khyber Pass. A punitive expedition restored the situation in 1842, penetrating deep into Afghanistan and burning parts of Kabul. These events virtually committed the Company to a policy of involuntary expansion and Sind was annexed the following year. In the Punjab, the warlike Sikhs feared that they would be the next to fall under the Company's domination, and decided to strike first. In 1845, a Sikh army swept across the Sutlej with Delhi as its objective. After four pitched battles fought at Mudki, Ferozshah, Aliwal and Sobraon, the invasion was decisively repulsed; in each the fighting possessed a ferocity which the British had not encountered since the Napoleonic Wars, and which came perilously close to breaking the Bengal Native Infantry. Consequently, much of the burden had fallen on the Queen's regiments and the Gurkhas.

A Council of Regency was imposed on the Punjab. This the Sikhs found hard to bear, but it was the very nature of their defeats that hurt them most. The *sahib-log*, the British officers, they could respect, for they had led as leaders should; they could also respect the *gora-log*, or white soldiers, who had fought hard and well. But the majority of their opponents had been Bengali sepoys, whom they despised as having neither instinct nor aptitude for war, and it was clear that they were thirsting for their revenge. Support was promised by the Afghans and, in 1848, the Punjab once more exploded into violence.

Thus far the 14th King's Light Dragoons had contributed squadrons to peripheral and comparatively minor affairs; now, the Regiment rode out knowing that it was about to fight some of its hardest battles. It could neither have known nor suspected that the first of those battles would become one of the major milestones in its history.

Ramnuggur

THROUGHOUT NOVEMBER 1848 both sides assembled their forces, the British at Ferozepore under General Lord Gough, and the Sikhs under Shere Singh along the Chenab river, near a town named Ramnuggur. Between lay a covering force commanded by General Sir Colin Campbell, the main element of which was a cavalry division of two brigades under the command of Brigadier-General C. R. Cureton, of which the Regiment formed part of the 1st Brigade, together with the 3rd Light Dragoons and the 5th and 8th Bengal Light Cavalry.

Cureton was a very remarkable man indeed and had risen from the ranks entirely by his own efforts. He was the son of respectable Lancashire tradespeople who had seen to it that he received a good education. During the Peninsular War he had served as an NCO in Brotherton's troop where his ability and intelligence had attracted such favourable attention that he was transferred to Staff duties and later granted a commission. Brotherton had been very reluctant to let him go, but did not stand in his way.

By 21st November, Gough's army was on the move and Campbell was ordered to carry out a reconnaissance in force, with the added task of clearing the Sikhs from the south bank of the Chenab. Like many Indian rivers, the Chenab is broad but rarely full, flowing through a number of winding channels but leaving wide areas of its bed apparently dry; this appearance is treacherous, for both the exposed areas of bed and the floodplain beside the banks consist of either deep, dry sand or a thin crust below which lay hock-deep mud. It was, in short, very bad cavalry country.

Contact with the enemy was made early on the morning of the 22nd. The 3rd Light Dragoons charged a body of cavalry which had crossed the river, driving it back. This action brought the leading British troops within range of the artillery which the Sikhs had massed on their own bank, and it was decided to retire a little to avoid needless loss. During this manoeuvre, one gun and two limbers belonging to the supporting Horse Artillery battery became stuck in the sand and had to be abandoned in the face of heavy fire. Jubilant, the Sikhs began swarming across the river to take possession of these trophies, and the Fourteenth were ordered in to restore the situation.

The Regiment, commanded by Lieutenant-Colonel William Havelock, was coming up in column of troops, and so far as can be ascertained its order of march was 1st Squadron, 2nd Squadron, 4th Squadron and 3rd Squadron; behind was the 5th Light Cavalry. Havelock, who had first seen action as a fifteen-year-old Ensign with the 43rd Light Infantry in the Peninsula, was one of those men who seem to find total release in battle and he had ridden eagerly towards the sound of the guns. Quite possibly the orders he received were unclear; or again, perhaps drifting dust and smoke prevented him from identifying the real objective. What is certain is that he identified a particularly large body of the enemy as his target and immediately swung the 1st and 2nd Squadrons into line, sending his adjutant, Lieutenant Richard Apthorp, galloping down the column with instructions for Lieutenant-Colonel John King that the 3rd Squadron should remain in reserve and act as a firm base upon which the Regiment could rally; no orders seem to have reached the 4th Squadron

The right flank of the 14th Light Dragoons' charge at Ramnuggur.

and it was at the initiative of its commander, Captain William Wilmer, that it conformed to Havelock's movements.

Gough himself had now arrived on the field and had joined Campbell and Cureton, and with fascinated horror he watched the Regiment attack the enemy where they were strongest. The Sikhs on the south bank promptly gave way before the charge and retired across a nullah to an island in mid-stream, which was held by approximately 3,000 men and a six-gun battery. Havelock, now in full cry, pursued them. The squadrons plunged over the steep river bank, thereby opening the distance between ranks, which became further disordered as the horses floundered through the heavy going. This proved to be a blessing in disguise, for the Regiment was now advancing into the teeth of the enemy's fire from the island, as well as that from the north bank of the river; much of this was aimed too high, but had the squadrons remained in close order they would certainly have suffered more grievously than they did. Although outnumbered by ten to one, when it came to handstrokes the Regiment proved that it was master of its craft; according to one eye-witness it sabred hundreds of the enemy. The arrival of Wilmer's squadron a little later than the others was probably the turning point of this phase of the engagement, and is treated in some accounts as a separate charge.

Suddenly, the Sikhs on the island broke and fled enabling the Regiment, which was presently joined by part of the 5th Light Cavalry, to rally and re-form ranks. Logic dictated that it should now retire but Havelock, who was clearly possessed of the joy of battle, had seen that the nullahs beyond the island were infested with Sikh infantry and was determined to clear them out. With his cavalryman's instinct Cureton read his mind and decided that enough was enough; he was riding forward to order a retirement when he was shot dead.

Once again the high notes of the *Charge* rang out and both regiments returned to the attack. The Sikhs fought bitterly but were pressed slowly back; of this, the bloodiest phase of the entire action, it is natural that no two survivors' accounts quite agree, but there are brilliant flashes of memory common to most. Of Havelock's grey going down and a knot of troopers closing instantly round him; his body, and nine of theirs, were found together after the battle. Of Captain Herbert Gall siezing a Sikh standard and having his right hand all but severed in the process; he never recovered its full use but, like Hervey, he chose to remain with the Regiment. Of Captain Wilmer and four troopers dismounting under intense fire to bring in the mortally wounded Captain Fitzgerald. Of Cornet D'Urban Blyth, a deft swordsman, riding to the aid of the wounded Lieutenant McMahon and killing four Sikhs in succession in single combat; with considerable *sangfroid* Apthorp told him only moments later that he was excused further sword drill!

Having completed their self-imposed task, both regiments withdrew to their own bank of the river. In total, the Fourteenth lost two officers, one sergeant, thirteen other ranks and thirty-seven horses killed, and five officers, four sergeants, eighteen other ranks and fifteen horses wounded. Considering the nature of the work, these figures are almost unbelievably low. The full extent of the Sikhs' casualties will never be known, but they exceeded those of the Regiment by many times; worst of all was the sharp blow to their morale, for it was inconceivable that so few should attack - and defeat - so many.

The Fourteenth's attack at Ramnuggur was, like that of the Light Brigade in the Crimea, undoubtedly a mistake, but its consequences were far from barren. A witness commented that 'All agree in one thing, however nobly the Fourteenth gained their laurels in the Peninsula, no charge they ever made could surpass this in gallantry, and yet no more than four or five of the officers and men had ever been under fire previously'. No battle honour was ever awarded, for in the Army's scale of actions the engagement was classed as being 'an affair of cavalry outposts'; on the other hand, the 5th Light Cavalry presented the Regiment with a magnificent silver cup to commemorate the event, and on every anniversary of the battle this is used to toast 'The Heroes of Ramnuggur'. Moreover, the implications of the action extend far beyond the physical and moral damage inflicted on the enemy – the lesson of Ramnuggur for successive generations within the Regiment is that once a task has been set it will be carried through, no matter how great the odds or however severe the physical difficulties might be.

The last survivor of the charge was Troop Sergeant-Major John Stratford, who died as recently as 1932, aged 103 years. It is an interesting reflection that older members of the Regimental Association may have met and talked with Stratford, who himself went into action at Ramnuggur under a Brigadier who had served in Brotherton's troop during the Peninsular War. There can be few better examples with which to illustrate the continuity of regimental life.

Troop Sergeant-Major John Stratford, the last survivor of the charge at Ramnuggur, photographed in his nineties. TSM Stratford died in 1932, aged 103.

Chillianwalla and Gujerat

FOR THE OFFICERS, the Christmas of 1848 was a boisterous affair, the most notable event being an affray in which plum puddings were used as missiles; from this Lieutenant-Colonel King, who had assumed command after Ramnuggur, emerged with a black eye.

Meanwhile, the war continued. Gough instituted a turning movement which forced the Sikhs to abandon their line on the Chenab, and then followed them up as far as Chillianwalla, where a pitched battle was fought on 13th January 1849. The issue was decided by a brutal infantry combat fought in close country, and the Sikhs retired from the field.

It was the Regiment's misfortune that it was brigaded, together with the 9th Lancers and some squadrons from the 1st and 6th Bengal Light Cavalry, under a Brigadier Pope, who was one of the Company's elderly generals. Pope was unable to mount his horse without assistance, and his grasp of cavalry tactics was similarly infirm. His brigade was posted on the right of Gough's line, and he instructed it to advance in one single line, without either scouts ahead or squadrons in reserve. During the advance the brigade drifted to the left, so that it began to mask the fire of its supporting horse artillery batteries. The Fourteenth formed the brigade's left wing and King very properly brought the situation to Pope's attention. The simple solution was for squadrons to take ground to the right, and the appropriate command was 'Threes – Right!' Inexplicably, Pope gave the order 'Threes – About!' The Light Cavalry squadrons in the centre wheeled to the rear. Puzzled, the British regiments on the outer wings conformed. Sensing weakness, the Sikh cavalry charged. Now thoroughly alarmed, the Light Cavalry dug in their spurs and were not seen for the rest of the day. A number of guns were over-run and while the British regiments maintained formation they were forced to gallop in order to gain sufficient ground in which to turn on their pursuers. Once this had been achieved the threat melted away, but it had been an impossibly difficult situation and, although a number of the enemy were killed, pointless casualties had been incurred. The Regiment lost Lieutenant A. J. Cureton – the General's son – and Private Tookey killed, and one officer and fourteen men wounded, two of whom later died. As for Brigadier Pope, his brief and troubled period of command was terminated by a sword cut to the head, but the stigma of having served in his brigade was to endure long after. The Fourteenth were extremely bitter about the affair and for many years to come the fists would fly in garrison canteens at the merest mention of the words 'Threes – About!'

After Chillianwalla, Shere Singh received substantial reinforcements, including a large Afghan contingent. He decided to move against Lahore and was about to cross the Chenab when Gough, who had also been reinforced, caught up with him. A further pitched battle was fought at Gujerat on 21st February, but by now the Sikhs had had enough and by mid-afternoon had been broken and were in headlong flight. The Regiment, serving with the 1st Light and 11th Irregular Cavalry under Brigadier G. H. Lockwood in the 2nd Cavalry Brigade, had spent the day on the right flank, balking potentially dangerous moves by large bodies of Afghan horsemen. It then took part in the general pursuit, which continued until dusk and which every account describes as being a thorough, ruthless affair. Amongst the trophies captured by the Fourteenth was a red silk standard, the bearer of which Corporal William Pain killed in single combat. The pursuit was resumed the following day, but the Sikh army had all but disintegrated. It remained only to harry the Afghans across the Punjab and through the Khyber Pass, and for the next few weeks the Regiment joined a flying column formed specifically for this purpose, finally halting at Fort Jumrud.

The Punjab was finally annexed on 1st April. One result of the mutual respect generated in two wars was that Sikhs in large numbers began entering British service. Their contribution to the history of the Empire and in both World Wars needs no further emphasis.

Lt-Colonel R. H. Gall. Born 1815, died 1887. Lt-Colonel Richard Herbert Gall, CB was gazetted Cornet on 3rd July 1835, became Lieut by purchase on 7th December 1838 and Captain on 31st March 1848. He served with the Fourteenth throughout the Punjab Campaign of 1848–9. He received a severe wound in the charge at Ramnuggur attempting to capture a standard and fought in the battles of Chillianwalla and Goojerat. He received the campaign medal with two clasps. He served throughout the Persian expedition of 1857 and commanded a wing of the 14th Light Dragoons throughout the Mutiny, having been promoted Major on 18th September 1857. He received the Brevet rank of Lt-Colonel and the CB along with numerous mention in despaches. He was promoted Colonel and ADC to the Queen in June 1870 and served for a time as Ass. Quarter Master General at Doner from 25th May 1872 – Dec. 1874.

Persia 1857

THE SECOND SIKH WAR led to the Regiment acquiring a second nickname, this, naturally enough, being The Ramnuggur Boys. They were also known, less officially, as the Pagri Wallahs, for on all save formal occasions they had sensibly abandoned their shakos and wore instead their round forage caps, wrapped in a loose *pagri* (turban) or a white cover, with a neck-cloth attached. They had also abandoned the wearing of the constricting stock and gloves. These fashions spread rapidly throughout the British garrison of India so that within ten years, to the horror of traditionalists, regiments could be seen going into action in their shirt sleeves, and doing all the better for it.

The Fourteenth were stationed at Meerut when the Crimean War broke out in 1854. Orders to embark were received but these were cancelled almost immediately because of increasing tension between Persia and Great Britain. Persian troops had occupied the town of Herat in Afghanistan, and since British policy was to preserve Afghanistan as a buffer state between India and continued Russian expansion in Central Asia, this could not be tolerated. The Persians were quite intractable and in 1857 an expeditionary force, of which the Regiment formed part, sailed from Bombay to the Gulf. Bushire Island was occupied and here most of the Regiment remained, although H Troop under Captain R. B. Prettejohn accompanied a further landing which penetrated the Shatt-el-Arab. The Persians fled whenever they were encountered and after their forts at Mohamra had been captured and blown up in March they abandoned their claims to Herat. The expedition then returned to India, the Fourteenth's only losses in this brief campaign being twenty-two horses, the majority from disease.

The Indian Mutiny 1857–1859

ABOUT THIS TIME a lower pattern of shako was issued and the tight coatee was replaced by a looser jacket braided across the chest, and it is in this dress that the Regiment is generally depicted during the Great Mutiny which wracked the Indian sub-continent for almost two years.

The origins of the Mutiny are too well known to require more than the briefest repetition here. There were many Indians who welcomed the order the British brought with them, but there were others, notably some of the Princes and the Brahmin priesthood, who resented the curtailment of their powers. Thus, while *thagi* (ritual murder by strangulation in honour of the goddess Kali) and *sati* (the burning of widows on their husbands' funeral pyres) might be regarded as grand old customs by some sections of the community, the British found them abhorrent and had stamped on them hard. Again, the Company's Bengal Army was riddled with caste, violation of which could imperil a sepoy's soul but which, hinted the Brahmin, was deliberate British policy. The event which triggered the explosion was the issue of a new cartridge rumoured to be greased with a combination of pig and cow fat. As the pig is held to be unclean by Muslims and the cow is sacred to Hindus, sepoys of both religions declined to accept it. It was particularly unfortunate that the issue was made during the centenniel of the battle of Plassey, for the ancient prophesy said that in that year the British would be driven from India forever. All the signs were that serious trouble was on the way, but the Company's generals, grown old among the sepoys, refused to believe that they could mutiny and failed to act decisively.

The Mutiny broke out at Meerut on 11th May and spread rapidly. Suddenly the 40,000 British troops in India found themselves fighting for survival. During this, the most tragic episode in the history of the Raj, both sides committed acts of which they have no reason to be proud. It was not that the British soldier is by nature a vindictive fighter – indeed, once the shooting has stopped he tends to forget very quickly; one thing, however, will turn him into an implacable, merciless killer, and that is the cold-blooded massacre of his womenfolk and children. Many Native regiments remained loyal and for them the conflict was, in effect, a civil war, with all the bitterness that that implied. This aspect of the matter was emphasised by the fact that mutinous regiments still wore their uniforms, still carried their colours in the place of honour, and were still played into action by their bands to the strains of the old English county airs taught them by long-forgotten bandmasters. Again, while some Indian prices, like the Nizam of Hyderabad, remained staunchly loyal to their treaties with the British, others, such as Nana Sahib, Rajah of Bitpur, and the Rani of Jhansi, rushed to join the rebels with their own state troops; the Rani, described as a handsome woman of twenty-four, dressed and fought like one of her own troopers, but despite this glamourous image was a sadistic enemy who mutilated her male British prisoners in the most obvious way.

The suppression of the rebellion was carried out in three phases. First, it was necessary to disarm or disperse regiments of dubious loyalty and, simultaneously, to relieve isolated garrisons which were besieged by the rebels. Secondly, once the outbreak had been contained, the rebels had to be decisively defeated in the field and their strongholds reduced. Finally, the survivors were hunted down and the ringleaders brought to trial and punished. Given the vast nature of the country, it was inevitable that these phases were executed at different times in different places, but for the British two factors were constant. The first was the use of mobility to regain the strategic initiative, and the second was the imposition of moral domination, achieved by attacking the enemy wherever he was to be found, irrespective of the odds. After the initial outbreaks, these methods ensured that it was the rebels who were forced to react to British moves, and not *vice versa.*

The Regiment, commanded by Colonel Charles Steuart, had barely settled into its new station at Kirkee on its return from Persia when it was required to take the field, and was divided into two wings which formed part of separate columns. The left wing, together with the 25th Native Infantry and an artillery battery, was ordered to relieve Mhow but was forced to divert to Aurungabad, where several mutinous regiments were dispersed without difficulty on 23rd June. The following day, a detachment under Captain Gall set out for Boldana, 100 miles distant, arrested the rebels there and returned to Aurungabad with their prisoners on 6th July. Continuing its advance, the column crossed the Nerbudda River on 27th July, was joined next day by the 3rd Hyderabad Cavalry, and entered Mhow on 1st August, where it was further reinforced by the arrival of four companies of the 86th Regiment, The Royal Irish Rifles.

The monsoon was now at its height and further operations were impossible until October. On 18th September, Gall was promoted Major, and since he was to play a prominent part in subsequent actions he deserves more than a passing mention. He had, it will be recalled, virtually lost the use of his right hand at Ramnuggur. After that, he had experimentally fitted a sword to his wrist but this device had not been successful and he now fought with his left hand. He was a small, spare, sallow man, driven by nervous energy, who was not liked by his men and who was actually disliked by some of his brother officers, probably because of his fussy, fastidious ways; he would, for example, sleep fully clothed and would often rise during the night to ride round the picquets. On the other hand he was universally admired for his courage, ability and horsemanship; when leading a charge he would turn in his saddle and challenge his men to overtake him – none ever succeeded.

In October, the column, now known as the Malwa Field Force, took the field again under Brigadier C. S. Stuart. Its objective was the rebel army of Feroz Shah, which outnumbered it by ten to one and was based at Mundesor. On the 22nd, a preliminary encounter took place at Dhar, in which Gall, Cornet Giles and a handful of troopers captured a three-gun battery, the guns being promptly turned on the enemy by the 25th Native Infantry. Further charges were made in which three standards were captured and the surviving rebels were pursued into a fort. This they later abandoned during the hours of darkness, leaving behind a quantity of treasure which was distributed as prize money.

Stuart's intention was to bypass Mundesor and relieve the besieged British community at Neemuch, twenty-two miles further on. His route, however, brought him close enough to Mundesor for the garrison to attempt a sortie against him on 21st November. That this was foiled was largely due to a twenty-strong light dragoon picquet commanded by Lieutenant George Dew. The enemy's advance guard numbered 300 infantry but, without hesitation, Dew and his men sailed into the mass, sabres flashing. Many of the rebels were cut down, others bowled over, and the rest bundled back into a river they had just crossed. Discouraged, the main body of the enemy retired. Dew was recommended for the Victoria Cross, but the award was not granted.

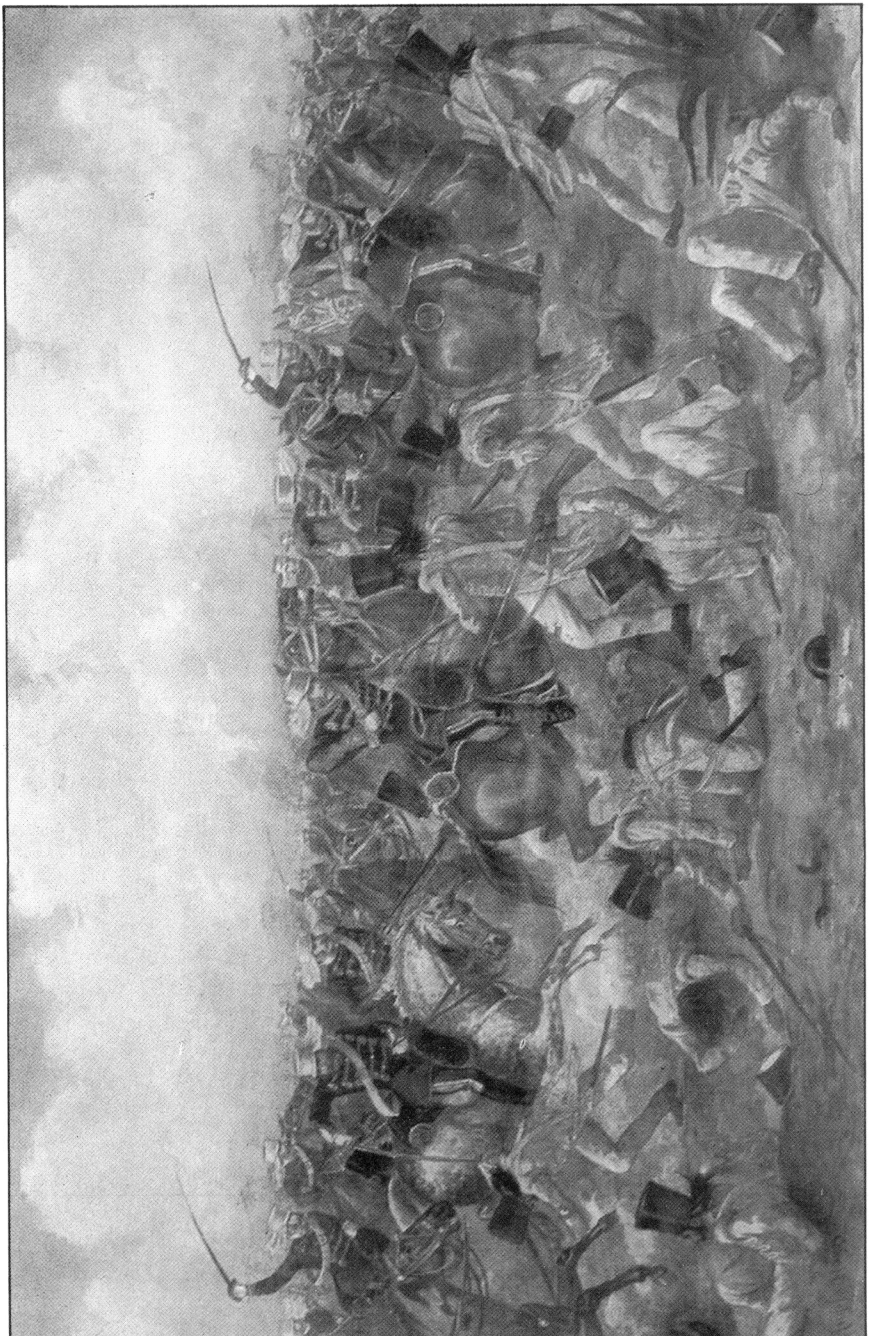

The Regiment in action against mutineers during the campaign in Central India.

The Mundesor garrison again attempted to interfere with Stuart's move the following day, during which the Fourteenth and the Hyderabad Cavalry were engaged in holding them off. An opportunity for offensive action occurred near the village of Peeplia, where about 100 of the enemy were sabred.

Now thoroughly alarmed, Feroz Shah had abandoned the siege of Neemuch to take up a position at Goraria, and here on 23rd November a full scale battle took place. Stuart's artillery and infantry were in process of winning the fire fight when nineteen troopers of the Fourteenth, led by Lieutenant Martin, actually charged the rebels' battery and took possession of the guns. Heavy fire from the enemy infantry drove them back and wounded Martin. Gall, however, was quick to capitalise on success and charged through the battery again with a complete squadron, breaking the rebel infantry behind and cutting down 200 of them in the pursuit. This proved to be the decisive point of the engagement. Most of the enemy withdrew, but a large number retained possession of Goraria where, on the 25th, they fought to the death when the village was stormed by the Royal Irish Rifles and the 25th Native Infantry.

The battle of Goraria had cost the Malwa Field Force sixty killed and wounded, the Regiment's losses amounting to approximately one quarter. Feroz Shah's army, on the other hand, had been reduced to the lot of broken fugitives upon whom even the country people preyed, while Mundesor and Neemuch had been cleared of the rebels. 'Much of the success in quelling this insurrection,' wrote Colonel Durand, Officiating Agent to the Governor-General of Central India, to Lord Canning shortly after the battle, 'is due to the judicious daring, the thorough gallantry with which, whenever opportunity offered, Major Gall, his officers and men, sought close conflict with the enemy - a bold one, who often fought most desperately'.

Having thus fulfilled his mission, Stuart retired to Mhow, where his command was re-designated the 1st Brigade of the Central India Field Force, then assembling under Major-General Sir Hugh Rose. The 2nd Brigade, under command of Colonel Charles Steuart, was already assembling at Sehore and included the right wing of the Regiment which, under Major Arthur Scudamore, had not been involved to the same degree as the left.

The Central India Field Force marched out on 6th January 1858, its strategic objective being the elimination of the rebels in Jhansi and Gwalior, who represented a threat to British counter-insurgency operations further north. The first task of the 2nd Brigade, which General Rose accompanied, was the relief of the garrison of Saugor, which had been under siege for seven months. Before this could be accomplished, however, the fortress of Rathgur would have to be reduced.

On 26th January, Rose's siege train began battering away at its walls and within three days had effected a breach. The rebel Rajah of Banpur now closed in on the brigade's rear but his followers immediately took to their heels when threatened by the Fourteenth and the 3rd Bombay Light Cavalry. The garrison of the fortress, witnessing the débâcle, crept away during the night but joined forces with the Rajah's men and took up a position behind the River Bina near Barodia. From this they were smartly ejected during an engagement on 30th January, in which three troops of the Regiment played a prominent part.

The road to Saugor, the successful defence of which had been possible because the 31st Native Infantry had remained true to its salt, was now open. As the column arrived on 3rd February it was welcomed ecstatically by the ladies of the garrison: 'I will not be positive that they did not even *cheer* us!' commented an officer in total disbelief. The *mores* of the time permitted besieged ladies to load and fire beside their menfolk, but for them to express approval in any way beyond the fluttering of kerchiefs was obviously regarded as being a little raucous!

After resting for a few days at Saugor, Rose resumed his march. The fort at Garrakota was bombarded and abandoned by its garrison after a token resistance. Their departure was discovered on the morning of 13th February and a flying column was formed to pursue the fugitives. This included two troops of the Fourteenth, commanded respectively by Captains J. R. Brown and Arthur Need, the latter a noted swordsman; altogether 100 of the enemy were killed, Need personally accounting for five.

Rose's objective was Jhansi, the epicentre of the rebellion in Central India and the scene of a particularly vile massacre of European women and children some nine months earlier. To get there, however, his 2nd Brigade would have to cross a range of hills through which there were only two passes, the Marhat and the Muddenpore. Major Scudamore, with one squadron, a battery and a small number of infantry, was ordered to execute a feint against the Marhat, which was held in strength by the Rajah of Banpur, while Rose fought his way through the Muddenpore, held by the Rajah of Shagahr. These tactics proved successful, the Muddenpore being stormed at the point of the bayonet by the Company's 3rd European Regiment and the Hyderabad infantry.

The effect of this victory was profound. Every fortified place, including Banpur, which lay between the pass and the city of Jhansi was abandoned. Rose promptly ordered Brigadier Stuart and the 1st Brigade to join him at the Rani's stronghold. The 1st Brigade had been advancing on a parallel axis to the west, the only serious opposition encountered being at Chanderi, which was stormed by the Royal Irish Rifles and the 25th Native Infantry on 17th March.

The investment of Jhansi was completed on 22nd March and it was immediately clear that the fortress would prove a very hard nut to crack. The citadel, of massive construction, towered on a granite rock above the city, which was itself girdled by strong and well designed masonry walls with excellent fields of fire. These defences mounted forty guns and were manned by a 12,000-strong garrison. It was true that the enemy's morale was shaky, but the Rani's ferocious reputation kept her men at their posts; further, she had already asked the Nana Sahib for help and he had promptly despatched his principal commander, Tantia Topi, with 22,000 men and twenty-eight guns, to break the siege.

The advance of Tantia Topi presented Rose with an acute problem. His total strength amounted to approximately half that of the rebel army, yet he dare not strip his siege lines lest the Rani's garrison fall on his rear; equally, he dare not retire, since this would be tantamount to a major defeat in the field. To divide his resources in the presence of the enemy was fraught with risks that another general might have found unacceptable, but Rose knew his men and assembled a force of 1,500 with which to meet Tantia Topi when the latter attacked on 1st April.

On the left of his line he placed a battery and Captain Prettejohn's squadron; in the centre were his heavy guns, the 3rd Europeans, the 24th Native Infantry and the Hyderabad infantry; and on the right were one troop under Captain Need, a troop of Hyderabad cavalry and a troop of Horse Artillery. As Tantia Topi's horde surged forward, covered by the fire of its numerous guns, Rose ordered his infantry to lie down in dead ground. Then, at the critical moment, he ordered his whole line to charge; this was the last thing the enemy expected of this tiny force, and its dramatic consequences are described by one of the Regiment's NCOs, G. C. Stent:

'It was a glorious sight to see them thundering along, headed by the General and Captain

National Army Museum

Lieutenant James Leith won the Victoria Cross during the fighting at Jhansi, going to the rescue of Captain Need, who was surrounded by rebels and fighting desperately for his life.

Prettejohn, the latter of whom was bareheaded, and who fought and shouted like a demon. One minute, and they were among the enemy, and all that was to be seen was a confused mass of flashing swords and bayonets, struggling men and horses, and hoarse shouts of rage. From this seething, struggling mass our men emerged victorious. The rebels were thoroughly routed in this charge, and turned and fled; were rallied, formed up again, to be again charged and routed; and yet again, only to undergo the same infliction, losing all their guns, and finally bolting in the greatest confusion, pursued by out men, who cut up great numbers of them, stopping only at the river Betwa from sheer exhaustion. Many of the enemy who escaped our swords were drowned in attempting to cross the river; the whole of the ground passed over by our men was strewed with their bodies, and at the lowest estimate it was calculated that 1,500 of them must have been slain, and no doubt the wounded were at least as many more.'

During the fighting, Need had become separated and found himself on difficult, rocky ground, surrounded by rebels. Despite his skill with the sword, his saddle, reins and clothing were slashed to ribbons by the enemy's tulwars, and he would certainly have been killed had not Lieutenant James Leith come to his rescue. Leith was awarded the Victoria Cross, the first in the Regiment's history.

Some miles downstream an even smaller force under Brigadier Stuart, including Lieutenant Giles' troop, had dealt similarly with a second enemy attempt to break through to Jhansi.

Tantia Topi covered his retreat by setting fire to the jungle, and this effectively ended the general pursuit. He was a clever commander with a keen sense of self-preservation and at the first hint of trouble he could be relied upon, heavily escorted, to lead a determined rush to the rear. Notwithstanding, the battle of the Betwa, which had cost the rebels their artillery, numerous standards and very serious casualties, remains an astonishing lesson in the superiority of quality over sheer mass, a lesson which is particularly appropriate at the present time. The Regiment's losses in the engagement amounted to five men and eleven horses killed, and twenty-five men and sixteen horses wounded.

The result of the battle reduced the garrison of Jhansi to despair, deepened by the creation of a breach in their walls. Once this had been stormed on 3rd April they fled, many being cut down by Rose's cavalry. The Rani, however, managed to escape and fight another day.

After the fall of Jhansi, Rose rested his troops for a while and was reinforced by several companies of the Highland Light Infantry. On 22nd April, he set off for Calpee, 103 miles to the north-east, this being Tantia Topi's principal arsenal. At Kunch, his advance guard, commanded by the newly promoted Brevet Lieutenant-Colonel Gall, found the rebel army present in strength and carried out a thorough reconnaissance of its positions. Rose decided on a converging attack by both brigades and this went in early on the morning of 7th May. In less than an hour the rebels were beaten and streaming along the Calpee road with Tantia Topi at their head. At first the withdrawal was orderly and was covered by a mutinied regiment which remained perfectly steady, firing and retiring by sections as it had been taught. Such men evoked admiration but little pity. Three troops under Captains William McMahon and Blyth rode over them after making a difficult charge across ploughed ground, killing them almost to a man; elsewhere, another section of the rearguard was cut up by Captain Prettejohn's troop. The enemy column now became a jostling mob which abandoned its guns - one of which was secured by Blyth - and was pursued ruthlessly for seven miles.

It was now the hot season of the year and this engagement was fought in a temperature of 110 degrees. Fatal cases of heatstroke were commonplace, particularly among the infantry, although the Regiment also suffered its share. Rose was compelled to allow his men twenty-four hours in which to recover.

During the resumed march an enemy ambush was detected by Gall's advance guard and avoided. Skirmishes were now a daily occurrence, but, by 21st May, Rose had reached the outskirts of Calpee, having been reinforced by the Connaught Rangers, who had been operating with another column beyond the Jumna.

Present in Calpee were Tantia Topi, the Rani of Jhansi, the Nana Sahib's nephew Rao Sahib, and the rebel Nawab of Banda, together with every man they could muster. Calpee itself offered an ideal opportunity for defence in depth and Rose would probably have incurred serious loss had he tried to fight his way into the town. This advantage the rebels completely wasted by attacking on 22nd May. They were held with difficulty and then decisively repulsed by a counter-attack. Led by Gall, the Regiment, the Hyderabad Cavalry and the Horse Artillery carried out a pursuit which lasted eight miles, cutting down fugitives and capturing guns, elephants and baggage. Calpee was occupied the following day.

The rebels fled to Gwalior and thither Rose pursued them through the unbearable heat, routing them through the Morar cantonments on 16th June. The next day he was joined by a column from the Rajputana Field Force and in fighting at Kotah-ki-Serai, four miles south of Gwalior, Lakshmi Bai, Rani of Jhansi, was killed. Some accounts suggest that she was shot during the pursuit, others that she was hit by a fragment of shell, others that she was cut down by a trooper of the 8th Hussars. The only two certainties are that she was dressed as a man, and that her body was burned by her followers that night.

Gwalior was stormed on 20th June and the following day all that was left of the rebel army in Central India, 12,000 men, attempted a stand at Jowra-Alipore. The stand was brief and during the pursuit the enemy lost all their remaining guns, elephants and equipment; Tantia Topi, as usual, displayed a very clean pair of heels. For conspicuous gallantry during this engagement, Private Novell of Prettejohn's troop was recommended for the Victoria Cross.

The Central India Field Force had now fulfilled its task and was broken up. In six months it had marched 1,000 miles through country which had varied between jungle and semi-desert, had crossed mountains and rivers, fought innumerable battles against odds, conducted sieges, had been constantly victorious, and had remorselessly stamped out the rebellion in the areas through which it had passed. Its progress is one of the most remarkable stories in the annals of the Army.

The Regiment was also dispersed among columns which scoured the country for rebel bands. As the year closed, the rebellion flared briefly again when Feroz Shah returned to the fray, but his army was roundly beaten at Ranode on 17th December by a column commanded by Brigadier-General Sir Robert Napier, of which Prettejohn's squadron formed part. Immediately after the battle Napier wrote to Lieutenant-Colonel Scudamore, then commanding the Regiment:

'This morning the enemy were beaten and pursued with slaughter for eight miles, your glorious Fourteenth going a mere handful into the mass without looking twice, Prettejohn leading, as you would expect.'*

After the battle, a subaltern of the Mahratta Horse came across Prettejohn sitting on the ground, looking thoroughly fed up. He had received a severe sabre cut across the left thigh but this worried him a great deal less than the fact that his horse had trotted off, taking with it his supply of cigars!

Order was not fully restored until 1859. After leading his pursuers a merry chase, Tantia Topi was betrayed by one of his former confederates in April and promptly hanged. The Nana Sahib, at whose door history has laid responsibility for the hideous Cawnpore massacre, sought refuge in Nepal, where he was reported to have died in October; the report was unconfirmed and for a generation would-be Nana Sahibs kept popping up all over the sub-continent. His nephew Rao Sahib tried to evade detection by assuming the life of an ascetic in the jungles of the Punjab, but in 1862 he too was betrayed, tried and hanged. Feroz Shah survived until 1877, dying a pauper in Mecca.

The campaign had cost the Regiment one officer and 72 men killed. In recognition of their services during the Mutiny, Scudamore and Gall were nominated Companions of the Bath; Captains Prettejohn, Todd and Thompson became Brevet-Majors and subsequently Lieutenant-Colonels; and Captains Need and McMahon became Brevet-Majors. Regimental Sergeant-Major Thomas Clark, whose name appears constantly in despatches, was awarded a medal 'for gallant conduct in the field'.

The Regiment sailed for home in 1860, having been in India for 19 years; during this period it had earned the battle honours Chillianwallah, Goojerat, Punjab, Persia and Central India.

---

*Colonel Steuart, together with Leith and several others, was ordered home on sick leave after Kunch, and Napier had taken over the 2nd Brigade. Brevet Lieutenant-Colonel Scudamore subsequently commanded the Regiment, save for a short period during the Calpee operations, when he was prostrated by heatstroke; during this Gall had assumed temporary command.

Captain Prettejohn leading his squadron's charge at Ranode.

The Years of Peace

FORTY YEARS WERE to pass before the Fourteenth saw action again. In 1862, the Regiment was based at Newbridge, near Dublin, when it converted to hussars, the principal change in uniform being the replacement of the shako by a busby with a yellow bag and white plume. Soldiers are somewhat conservative by nature and many, particularly retired members of the Regiment, would have questioned sharply the necessity for change.* The older serving hands expressed serious doubts as to whether the busby provided the same protection as the shako, while the newly joined regretted that the very item which gave the hussar his characteristic swagger, the pelise, had been abandoned by the British Army some years earlier. Be all that as it may, the 14th (King's) Hussars were a fact of life and soon recognised as such.

The Regiment was stationed in various parts of the United Kingdom until 1876 and was for a time quartered in Lancashire, where the cotton famine induced by the American Civil War had resulted in widespread unemployment and the possibility of civil unrest. In 1876, it embarked for India, where it was stationed at Bangalore. To its intense disappointment, it did not take part in the Second Afghan War, but in 1881 it sailed for South Africa, where hostilities had broken out with the Transvaal Boers. However, the fighting was over by the time the Regiment arrived and after remaining some eight months it returned to India, where it was quartered at Secunderabad. In 1886, it embarked for home again and was serving in Ireland when the Second Boer War broke out in 1899.

During these forty years the basic nature of war changed more radically than it had throughout the previous three centuries. The muzzle-loading musket had been replaced by the magazine rifle, and the cannon by the quick-firing field gun; a new weapon, the machine-gun, had taken its place on the battlefield; and there had been a tremendous advance in ammunition technology and artillery techniques. This huge increase in fire-power meant that close-ordered ranks were now a thing of the past. For their own protection, soldiers began to dig in, using barbed wire to prevent attackers reaching their position. To many it was clear that the power of the defence had become stronger than that of the attack, and for the cavalry, in particular, the implications required the most serious and professional consideration.

*Traditionally, Light Dragoons had always looked askance at other branches of the Cavalry, describing their Heavier Brethren as 'Overweight Louts' and Hussars as 'Theatrical Bounders'.

The Second Boer War

THE DISCOVERY of mineral wealth inside the Boer republics of the Transvaal and the Orange Free State led to an influx of *Uitlanders*, as foreigners were termed. Most of these men and their families were British, but despite the fact that they produced the very wealth upon which the republics depended, they were denied the basic rights of citizenship and treated extremely badly. The Boers rejected all overtures on the subject by the British government and made no secret of the fact that they intended in due course to take over Natal and Cape Colony, so bringing the whole of southern Africa under their domination. The fumbled opening moves of the Zulu War and their own easy success in 1881 had made them contemptuous of the British Army and in October 1899 they declared war on the United Kingdom.

The Boers were able to mobilise 60,000 men, all trained shots and good horsemen, organised in locally raised Commandos of varying size. They also possessed a small but efficient artillery arm, manned by regulars and trained by German instructors; unlike the Royal Artillery, which still fought its batteries wheel to wheel, this was adept at dispersing and concealing its guns. Their principal assets, however, lay in their total mobility and their lack of dependence upon a supply line in the conventional sense.

At the start of the war there were comparatively few British troops in South Africa, the majority being widely dispersed infantry brigades, and these were quickly boxed in and besieged at Ladysmith, Kimberley and Mafeking. The next few months, therefore, were spent in trying to break the Boer hold on these towns. At home the public was a little surprised by this turn of events, but remained convinced that all would come right in the end. In fact, the sheer size of the landscape in which the conflict was being fought was not fully appreciated, and it took the General Staff a little time to decide, correctly, that the key to success lay in mobility and that in the final analysis the side which lost it would also lose the war.

A large number of mounted infantry battalions were raised, yet although these fully vindicated their creation, it was of the cavalry regiments that the Boers were most afraid. It was relatively easy for a Commando to break through a screen of mounted infantry, who were armed solely with the rifle and bayonet; no such liberties could be taken with cavalry, against whose sabres and lances the Boers had no defence at close quarters.

The Regiment, commanded by Lt-Colonel G. C. H. Hamilton, was at Newbridge when it was ordered to mobilise on 9th October 1899, but sickness among the horses prevented its immediate despatch to the theatre of war. It was not until 13th December that the main body embarked at Southampton, B Squadron following ten days later. While the Regiment was at sea it received the news of Black Week, in which those British troops already fighting in South Africa were sharply defeated in quick succession at Stormberg (10th December), Magersfontein (10th–11th December) and Colenso (15th December). One result of this was that General Sir Redvers Buller, whose operations had been dogged by a combination of bad luck and bad judgement, was replaced as Commander-in-Chief by Field Marshal Lord Roberts, and it was largely due to the latter's clear-sighted appreciation of the position, and that of his Chief of Staff, General Lord Kitchener, that the overall situation was so quickly brought under control.

The campaign on which the Regiment was about to embark was quite different from anything it had undertaken before; it had not fought a European enemy since the Napoleonic Wars, and this particular foe was a highly motivated expert in fieldcraft who, for much of the time, chose the time and place of engagements to suit himself. Initially the Regiment was organised in three sabre squadrons of four troops each, but a fourth squadron was added in June 1900 and maintained until February 1902. For most of the war this organisation remained theoretical, as a general shortage of water led to heavy losses among the horses, so that squadrons tended to be well below strength after the first few months in the field. To begin with, the Regiment was armed with the Lee-Metford Carbine, the performance of which compared most unfavourably with that of the Boer Mauser, but in November 1900 this was exchanged for the standard infantry rifle and engagements took place regularly at ranges up to 2,000 yards. Regarding tactics, 'Up to the capture of Pretoria the Regiment were comparatively strong in men and horses and all movements, attacks, dismounted action, outposts and protective duties were carried out as laid down in the training manuals which were in use to the commencement of the Great War. When moving over open country the formation was generally *Mass* or *Line of Squadron Columns*, the squadrons at half-open files for convenience and coolness in marching. The cavalry fighting throughout the war was mostly dismounted and generally by sections with mounted supports; there was an interval of 20 yards between men in dismounted action. In the attack threats to one or both flanks of the enemy's position were made, as the Boers were particularly nervous about being outflanked. Owing to the open undulating country Advanced Guards were kept well forward and all hills were occupied if possible before the Main Body came within effective artillery range of them. Rear and Flank Guards were similarly kept well out to keep the enemy out of rifle range of the column. The Advanced Guard kept its base 1½ miles in front of the leading brigade; entrenched piquets were used with mounted patrols sent out before dawn.'*

B Squadron disembarked at Cape Town on 10th January and moved up-country by stages to join Major-General John French's Cavalry Division, being attached for the moment to the Carabiniers. French's Division formed part of General Lord Methuen's army which was about to embark upon another attempt to relieve Kimberley. Methuen was detested by his men, who felt that his unimaginative tactics at Magersfontein and elsewhere had caused them quite needless loss. Fortunately, the projected operation would take place under the direct supervision of the Commander-in-Chief himself and 'Bobs', as he was known, possessed a strategic insight which places him firmly in the foremost rank of British generals.

The essence of Roberts' plan was a holding action designed to tie down the Boers in their trenches at Magersfontein and distract their attention from a wide flanking movement to the east of their position made by the rest of the army. This would be led by French's cavalry operating *en masse* and using their mobility to break through quickly to Kimberley; equally important was the fact that the Boer army at Magersfontein would be isolated from its base at Bloemfontein, the capital of the Orange Free State. In every respect, save the obvious lack of air support, the plan exactly paralleled the modern blitzkrieg technique.

---

*Notes prepared by Colonel Browne-Synge-Hutchinson, VC, CB, for the *Historical Record of the 14th (King's) Hussars*, Vol 2. At this period cavalry regiments were also equipped with one Maxim machine-gun, which was carried on a mule.

Addressing the senior officers of the Cavalry Division on 10th February, Roberts said: 'I am going to give you some very hard work to do, but at the same time you are going to get the greatest chance cavalry has ever had. You will remember what you are going to do all your lives, and when you have grown to be old men you will tell the story of the relief of Kimberley.'

The fords across the Riet and Modder rivers were siezed on the 14th and the final stage of the flank march began the following day. The movement of thousands of men and horses, 100 guns and hundreds of waggons could hardly have been kept a secret for long, despite an elaborate deception plan, but General Piet Cronje, the Boer commander, imagined that the manoeuvre was a feint designed to lure him out of his trenches at Magersfontein, against which he thought the main weight of the British attack would be directed, and sent only 800 men to confront French. This small force erected a wire fence at Klip Drift and then positioned themselves on kopjes surrounding a broad valley through which the Cavalry Division had to pass. The far end of the valley was closed by a low ridge and it was confidently expected by both sides that the clip-loading Mausers would do considerable execution among the packed ranks of the squadrons.

The wire was cut by a 25-strong forlorn-hope party provided by the 16th Lancers, most of whom were hit in the process. French then gave the order for the Division to advance up the valley beyond in column of brigades. At first the movement was executed at the trot, but the pace was increased to a gallop when the leading regiments began to come under fire at a range of 1,000–1,500 yards. Soon the entire formation was smothered in a dust cloud of its own making, and this in itself proved an adequate defence against the enemy's legendary marksmanship. The sheer moral effect of mass, accompanied by the earth-shaking thunder of more than 6,000 sets of shod hoofs, was too much for the Boers. Those lining the ridge fled, a number who left it too late being cut down or speared, and then the Cavalry Division was through, having accomplished what the manuals said could no longer be done, at the astonishingly low cost of only fifteen personnel casualties and twenty horses killed or injured.

Five miles beyond, French pulled up to reorganise and allow the horses to recover. Kimberley now lay only a few miles distant and the Boers hastily abandoned their siege lines. The advance was resumed with B Squadron 14th Hussars in the lead and it was actually Lieutenant Walker's troop which reached the town first.

Back at Magersfontein, the news left Cronje stunned. For several priceless hours he remained incapable of reaching any decision which would extract his command from the dangerous situation in which it was now placed. At length he decided to retire on Bloemfontein, passing through the gap which existed between the Cavalry Division and the slower infantry formations which were following up. The dust raised by his column was, however, spotted and he was run to ground at Paardeberg and again surrounded. As an iron ring closed around him, various Commandos rode to his aid and for the next ten days the Cavalry Division was employed in holding them off. On 27th February – Majuba Day in the Boer calendar – Cronje and more than 4,000 of his burghers surrendered. It was the worst defeat the Boers had ever sustained and, with the exception of Mafeking, which maintained a spirited defence until relieved on 17th May, virtually ended coherent warfare on the Western Front.

Roberts' next objective was Bloemfontein, and the Boers were in no position to stop him, although they did fight rearguard actions at Poplar Grove and Driefontein, B Squadron being involved in both. Bloemfontein was entered unopposed on 13th March, the troops being welcomed by a large section of the population; enthusiasm for the war had never been as great in the Orange Free State as it had in the neighbouring Transvaal.

The remainder of 14th Hussars had landed at Durban on 6th January and then moved up to the Tugela where Buller still retained local command and was still trying desperately to break through to Ladysmith. Two troops were present during the battle of Spion Kop on the 21st and came under fire there, but the next few weeks were to be spent by the Regiment in skirmishing with the Boers on the right flank of the army, where it quickly established a reputation for steadiness in action, despite the fact that only Colonel Hamilton and his Second-in-Command, Major E. D. J. O'Brien, had seen active service before, and that in the Second Afghan War, twenty years previously.

It was on the right flank that Buller eventually succeeded in breaking the Boer line during the battles of Vaal Krantz (5th–7th February) and Pieter's Hill (22nd February). These were largely infantry actions won at heavy cost, concerning which Buller received further unfavourable publicity. Nonetheless, by first light on the 28th it was evident that the Boers were pulling out, their decision to abandon the siege no doubt being further influenced by the news of Cronje's surrender. They were shadowed by A Squadron 14th Hussars under Major O'Brien and A Squadron 13th Hussars, both of which had had it firmly impressed upon them that they should not incur casualties, an issue on which Buller had become extremely sensitive. O'Brien remained convinced that a major opportunity existed at this point for the army's cavalry brigade, which was fresh and eager to turn the Boer defeat into a rout, and that this was lost as a result of Buller's hesitation.

Ladysmith was relieved the same day, its garrison on the verge of starvation. Some people were supplementing their meagre rations with such items as dubbin and birdseed; real food, when available, fetched absurd prices. A dozen eggs, for example, cost £2.8s.0d. and 1 lb of beef fat 11s.0d. Smokers were particularly hard hit, a packet of cigarettes costing £1.5.0d, and ¼ lb of tobacco £2.5s.0d., and a dozen matches 13s.6d.

On conclusion of the Ladysmith operations, the Regiment was brought round by sea and train to Bloemfontein, where it joined the 4th Cavalry Brigade (7th Dragoon Guards, 8th and 14th Hussars and O Battery RHA) under Major-General J. B. B. Dickson; B Squadron rejoined at Donkershoek on 14th April.

The Boer Commandos had now become active again and had not only effected the temporary capture of Sanna's Post but also bottled up a British detachment at Wepener. On 21st April the 4th Cavalry Brigade was detailed, with other troops, to effect the relief of this, which it successfully accomplished after fighting a number of small-scale actions at Leuwkop, Paardekraal and Roodekop. The last involved the Regiment in a fierce fire-fight lasting twenty minutes in which it sustained the heaviest casualties of any engagement throughout the war – one officer and two men killed, and eight men wounded. The enemy was compelled to abandon the siege, but on the 28th launched a counter-attack at Thabanchu Mountain which was intended to encircle the brigade. They were thwarted by a skilful withdrawal in difficult circumstances, with the Regiment acting as rearguard during the most awkward phase of the battle.

After the relief of Wepener, the Cavalry Division as a whole took part in Roberts' resumed advance, which was made northwards along the railway through Kroonstad to the capital of the Transvaal, Pretoria. The Regiment was slightly involved in the battle of Zand River on 10th May, played an important part in the siezure of the strategically important Valsch River Drift ahead of the Boers on the 11th, having covered over 50 miles during the day, and entered Kroonstad on the 12th, where a halt was called for ten days. During the final phase of the advance the Fourteenth was in almost daily contact with the enemy and took part in the actions at Vlakfontein, Klipriversberg, Dornkop and Orange Grove, in which the Cavalry Division was repeatedly used to lever the Boers out of their position by threatening their flank. Johannesburg fell on 31st May and Pretoria four days later.

Here the Regiment acquired one of its more unusual recruits, a baboon named Kruger who bore more than a passing resemblance to his namesake, the President of the Transvaal Republic. He evidently liked the military lifestyle, for he remained on strength until 1905. Like most soldiers, he enjoyed a good binge now and again, his favourite tipple being lamp oil, which did not agree with him.

The real Mr Kruger had meanwhile retired along the railway running east from Pretoria to Portuguese East Africa, although some of his troops remained in the vicinity of the capital and a fierce engagement took place at Diamond Hill on 12th June, in which the Regiment successfully held an exposed feature during two days of close-range fighting. Having completed his mopping up, Roberts began what he hoped would be the final phase of the war by advancing along the railway in pursuit of the president and the remains of the Boer field army. The Regiment took part in the actions of Waterpan, Belfast, Silverkop, Nelshoek and the capture of Barberton on 13th September. Mr Kruger had crossed the frontier two days earlier and was to embark aboard the Dutch cruiser *Gelderland* in a vain attempt to rally practical support for the Boer cause throughout Europe.

The war seemed to be over, in the conventional sense. Many Boers, however, could accept neither the annexation of their republics nor the oath of loyalty to the Queen, and these decided to conduct a guerrilla campaign, hitting the British hard wherever and whenever possible. The first effects of this were felt by the Regiment on 13th October at Geluk Farm, near Belfast, where a small column of which it formed part, together with the 8th Hussars and M Battery RHA, was attacked in its camp at dawn by a large and very determined Commando. The Boers made such good use of ground that they were soon threatening both flanks and inflicting mounting loss on the 8th Hussars and the Gunners. The column commander decided to pull out before the enemy closed the one remaining line of retreat and detailed the Fourteenth to act as rearguard.

Colonel Hamilton asked Major E. D. Brown to co-ordinate the movements of A and C Squadrons while they broke contact. This was extremely difficult as the enemy's fire had unsettled the led horses and they were starting to break loose. Brown personally caught three for their riders and then took up Sergeant Hersey behind him when the latter's mount was shot dead. Having lifted Hersey to safety, Brown returned to look for more dismounted stragglers, accompanied by his trumpeter, Lance-Corporal Leigh. They found Private Gregory, a heavy man who could not be mounted double, so Leigh handed him his own horse and jumped up behind Brown. Nearby, another man was brought in by Hersey, who had obtained a fresh mount. The withdrawal continued by bounds, with the Boers pressing hard and threatening to cut off the better part of the two squadrons. At one phase Lieutenant J. G. Browne (subsequently the 14th Hussars' last Commanding Officer) was having great difficulty in mounting his charger, which was wild with excitement and on the point of bolting, but Brown galloped across and held the animal's head until it could be brought under control. At length the rearguard reached the main body of the column, the fire of which finally drove off the attackers.

It was later established that the Boers, commanded by General Tobias Smuts, belonged to the Ermelo, Carolina and Heidelberg Commandos and were approximately 1,000 strong. They had come dangerously close to destroying the column but had been foiled by

Portrait of Colonel Browne-Synge-Hutchinson, VC.

the Regiment's steadiness and tactical sense. As it was, they had inflicted some seventy casualties, the Fourteenth losing only two killed and eight wounded, which was remarkably light considering the extent to which the Regiment had been involved. Several men owed their lives directly to Brown, whose gallant conduct was recognised by the award of the Victoria Cross.*

On 21st November, Colonel Hamilton was appointed commander of the 4th Cavalry Brigade and temporary command of the Regiment was assumed by Brevet Lt-Colonel Brown, VC, in the absence of Brevet Lt-Colonel O'Brien, who had returned to England after the engagement at Diamond Hill; O'Brien took command on re-joining in June 1901.

The Boers' guerrilla tactics were proving remarkably effective against conventional

*In 1904 he changed his name to Browne by Royal Licence and assumed the additional surnames of Synge-Hutchinson, these being his mother's family names.

troops who could not be strong everywhere. On 24th December, for example, Lieutenant the Hon H. Grosvenor's 25-strong patrol was involved in a six-mile running fight with a much stronger party of Boers, losing thirteen men killed, wounded or temporarily captured by the enemy. Grosvenor and Private Shenton were both commended for their gallantry in bringing in dismounted men, and Shenton was awarded the DCM.

Lord Kitchener had replaced Roberts as Commander-in-Chief when the latter went home in November, and he was well aware that there was no quick solution to the problem. Many Boer prisoners had been released on parole but had broken their word and taken the field again, and the Boer farming community as a whole was providing support for the Commandos. It was, therefore, reluctantly decided that this infrastructure must be destroyed if the war was to be ended. This involved moving families into camps far removed from the war zone and, in some cases, burning farms which were known Commando refuges. Kitchener also decided to inhibit the Commandos' mobility by building blockhouse stop-lines which divided the country into smaller areas which could be combed by mounted columns on the cordon-and-search principle. These measures caused an unavoidable degree of hardship to the civilian population but the very nature of the war made them inevitable. Many historians suggest that the concept as a whole was British in origin; in fact, the Spanish Army had been forced to employ identical measures against equally intransigent Cuban *insurrectos* only a few years previously.

In pursuance of this policy the Regiment was split up at the end of December 1900 and did not serve together again until February 1902. During this period the squadrons were engaged with the Boers on nine separate occasions. Of these the most significant was an action fought near Hartebeestefontein on 24th March 1901, involving the column under General J. M. Babington of which A Squadron formed part. The enemy on this occasion were commanded by General Koos De la Rey, one of the most capable and energetic of the Boer officers, who had assembled a large force for an attack on the garrison of Lichtenburg, which had been repulsed only with difficulty. It was while he was retiring from this foray that Babington's column caught up with him and promptly made a mounted attack. The Boers fled in all directions, thirty being killed and 140 captured, the majority belonging to Smuts' Commando, with whom the Fourteenth had been so closely engaged at Geluk Farm. In a pursuit which lasted for twelve miles, De la Rey also lost his entire waggon train of fifty-three vehicles, two field guns which had been captured during the Colenso disaster, one pom-pom, and five Maxim machine-guns.

It was actions such as this which, together with drives between the blockhouse lines, finally eroded the Boers' strength and their will to continue fighting, although they remained extremely dangerous opponents to the very end. The last major operation in which the Regiment took part was the celebrated Harrismith Drive of February 1902 which, together with its results, is described elsewhere in this book.

An honourable peace treaty was concluded in May 1902 and the Fourteenth went into garrison at Kroonstad. Losses in action for the whole campaign amounted to one officer and seventeen other ranks killed, and two officers and forty-two other ranks wounded.

In March 1903, the Regiment embarked for home, having first sold its troop horses to the Boers as part of the government's wise policy of reconciliation. Its march from Aldershot station to barracks was therefore made dismounted; young Kruger, the baboon, thoroughly enjoyed the event, capering about behind the band with his minder and generally up-staging everyone to the delight of the crowd.

Home from the war. Kruger, the Regiment's baboon mascot, contemplates the wonders of Aldershot.

Towards World War

THE REGIMENT embarked for a further tour of duty in India in 1906. During this period, and indeed right up to the outbreak of the First World War, the lessons of the Second Boer War were being thoroughly digested, the result being that in 1914 the British Army was, for its size, the most formidable and professional to take the field.

For the cavalry it was particularly appreciated that firepower would play an increasingly important element in its operations. The Short Lee-Enfield rifle, the same weapon used by the infantry, was issued and great emphasis was placed on sustained, rapid and accurate fire. In 1911, the regimental establishment of four sabre squadrons was augmented by the addition of a two-gun Maxim machine-gun section. The *arme blanche*, however, was not forgotten, the sabre being replaced by the same pattern of straight sword used on ceremonial occasions today, the advantage being that when extended beyond the horse's head the latter had almost the same reach as the lance.

Individual effort directed to the common good having always formed part of the Regiment's tradition, in 1909, Lieutenant-Colonel Browne-Synge-Hutchinson, VC, re-introduced a custom which had long fallen into abeyance, that of awarding a Regimental Medal 'to any officer, warrant officer, non-commissioned officer or man of the Regiment who contributed in some conspicuous manner to the military efficiency or the military honour of the Regiment'. Further details concerning the award, and a list of its recipients, can be found in Appendix F.

When war was declared in August 1914, the Regiment was serving in Meerut, but ironically it was the few men manning its Depot in the now empty cavalry barracks at Scarborough who came under fire first. The misty dawn of 16th December revealed the battle-cruisers *Derfflinger* and *Von Der Tann* lying offshore and these opened a heavy and indiscriminate fire into the undefended resort. There were no casualties in the barracks but many civilians were killed and injured and there was widespread destruction of property.

This sort of incident contributed to the intense hatred of all things German that was then sweeping the Empire. At a mess meeting held in May 1915, it was resolved that the Regiment's second badge, the Royal Crest within the Garter, should replace the Prussian Eagle, the wearing of which was discontinued for the time being.

The order to mobilise for active service was received on 27th October 1915 and the following month the Regiment embarked at Karachi for Mesopotamia, which occupied much the same area as modern Iraq, but which then formed a province of the Turkish Empire.

A fine study of the Regiment's Warrant Officers and Sergeants, taken at Shorncliffe in 1906.

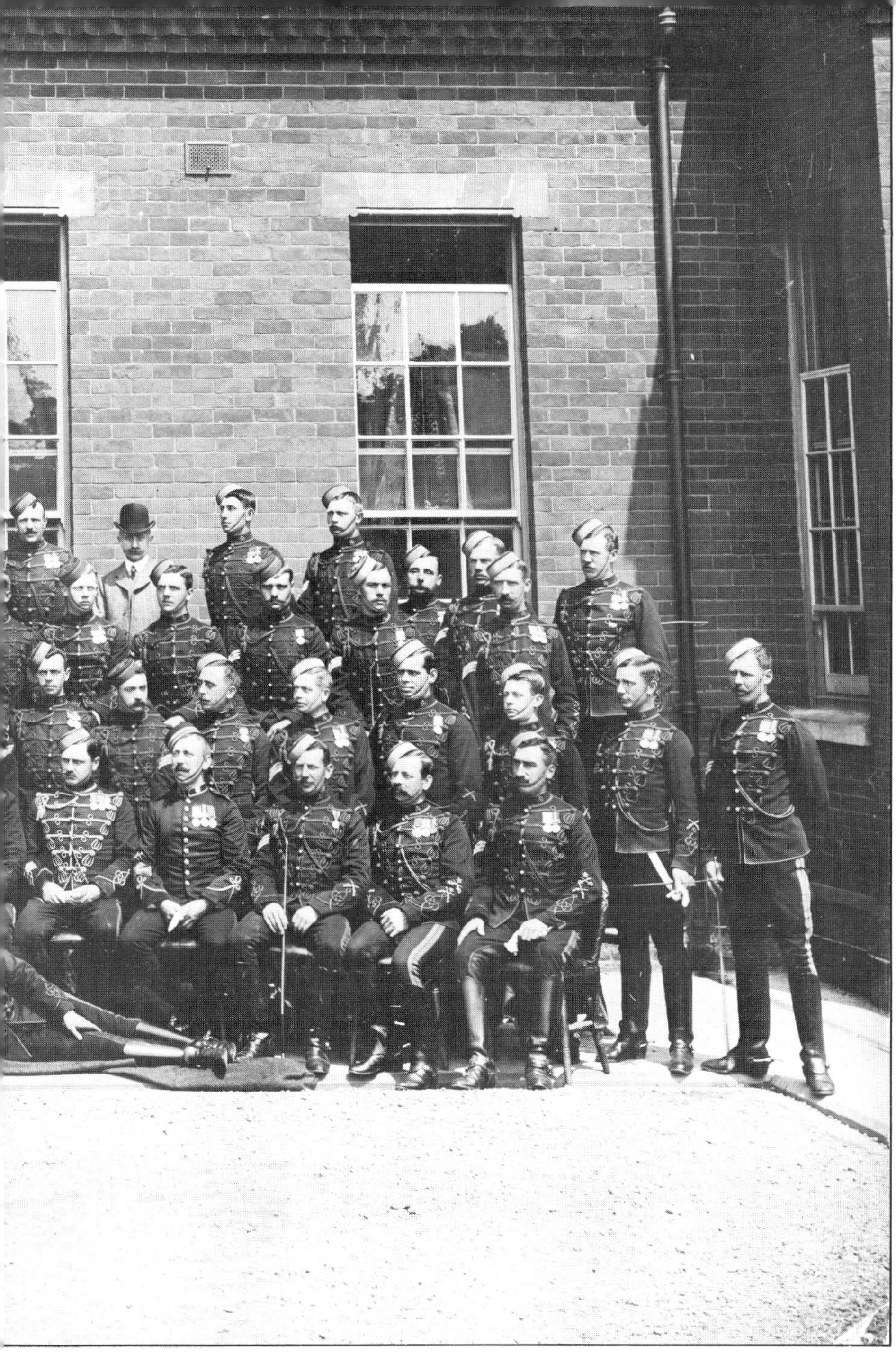

Major T. Gossage

The Kut-al-Amara Campaign

MESOPOTAMIA IS AN unpleasant country in which to conduct a campaign, but the necessity of doing so was occasioned by the need to secure those oilfields close to the Persian Gulf and simultaneously deny their use to the Central Powers, with whom Turkey was allied. Of Lower Mesopotamia in particular, an earlier historian of the Regiment commented: 'This great delta of the Tigris and Euphrates is peculiarly lacking in natural features. The great marshes encountered were generally too saline and brackish for the use of man and horse, and as there was scarcely any other water away from the big rivers, military operations were considerably limited in consequence, and elaborate and prolonged outflanking movements were to a great extent unfeasible. In the great area between Basra and Daur, apart from Baghdad and a few riverine village gardens, there was not a tree. This lack of landmarks necessitated all moves being made by compass and affected tactics to the extend that there were no features to make the objective or bound, and there was no cover to provide shelter from view. Minor tactics were in consequence particularly simple. There was little opportunity of making use of ground and units advanced from one point to another in a direct line. Lack of cover increased vulnerability and minimized the chance of effecting surprise; consequently, approach marches by night increased both in length and frequency.'

In summer, the sun blazed down on this wilderness with furnace intensity; in winter, incessant rain created universal mud in which no rest was possible. By day flies crawled over and into everything, by night clouds of mosquitoes arose to torment, sting and infect, only sandstorms providing some temporary relief from these pests. Together, cholera and dysentery killed almost as many men as the Turks. To these natural hazards was added an especially venomous species of Arab scavenger who hovered around the armies, in view but generally out of range, ready to murder stragglers, butcher the wounded and then vanish into the desert with his pickings.

The Turkish Army had recently been re-organised and re-trained with German assistance but had still not digested the full implications of twentieth-century warfare. German staff officers serving with the Turks were sometimes driven to distraction by the latter's reluctance to replace the liesurely fatalism of Islam with sound planning, particularly in the field of logistics. Notwithstanding, the Turkish soldier himself was a hardy, unimaginative and extremely brave individual who was an opponent to respect.

Operations in Mesopotamia had begun a year earlier with the siezure of the Shatt-al-Arab waterway and Basra. Turkish counter-attacks had been defeated with such comparative ease that the government of India, responsible for this theatre of war, had assumed that the poor quality of the troops encountered thus far was representative of the Turkish Army as a whole, and had now decided that an advance should be made on Baghdad, the loss of which would seriously damage the enemy's prestige throughout the Muslim world.

The British forces in Mesopotamia at this period were commanded by General Sir John Nixon and consisted of two infantry divisions and a cavalry brigade. Nixon had enjoyed a long and honourable career in the Indian Army but has been described by his critics as being impetuous, needlessly optimistic and lacking in strategic sense.

The commander of Nixon's leading formation, the 6th Indian Division, was Major-General Sir Charles Townshend, who had first attracted public attention during the epic defence of Chitral Fort in 1895 and gone on to serve in the Dervish wars. Undeniably courageous and professional, Townshend did possess an acute sense of strategy, but this he was prepared to subordinate to his own personal interests; narcissistic by nature, he was consumed by ungovernable ambition and sought preferment and publicity wherever they might be found. Together, the combination of Nixon and Townshend boded ill for those who served under them.

For a while all went well. Accompanied by a flotilla of gunboats, Townshend advanced up the Tigris and captured Amara on 3rd June. After pausing to reorganise, he resumed his advance and inflicted a serious defeat on the Turks at Kut on 27th/28th September. Despite the fact that he was becoming dangerously over-extended he was ordered to continue his march on Baghdad. By now, however, regular Turkish divisions were arriving in Mesopotamia and had constructed a well-fortified position at Ctesiphon. This, Townshend attacked on 22nd November, but four days of intense fighting failed to achieve a complete breakthrough, despite 6,200 casualties inflicted on the enemy. Moreover, British losses amounted to 4,600 out of a total of 11,000 and, as the Turks were being steadily reinforced as the battle progressed, Townsend decided that he must withdraw.

The 14th Hussars, commanded by Major R. W. Hewitt, had disembarked at Kut on 23rd November and reached the retreating army at Aziziyeh five days later. Here the Regiment joined the 6th Cavalry Brigade, which also included three Indian regiments (7th Lancers, 16th and 33rd Cavalry) and S Battery, Royal Horse Artillery.

Townshend's withdrawal was hampered by the large number of wounded requiring transport and by many of his river craft running aground on shoals. During the evening of the 28th, the gunboat *Shaitan* stuck fast eight miles from Aziziyeh, springing her plates. It was decided to abandon her but while her guns and ammunition were being transferred she came under fire from Arab irregulars on the bank. The cavalry brigade was sent forward with the Fourteenth leading and soon large numbers of the enemy were seen around the village of El Kutunie. Covered by a dust storm, A Squadron drew swords and charged home; taken completely by surprise, the Arabs fled, only to find themselves trapped by the 7th Lancers, who had attacked from a different direction. Between them, the two regiments cut down about 140 and shot many more who sought to escape by swimming. The affray was witnessed by the approaching Turkish advance guard, which was promptly shelled by S Battery and brought to a standstill. After this, the Arabs remained very much in awe of the cavalry and made no further attempt to interfere directly in operations.

The retreat continued, but progress was so slow that by 1st December the Turks had caught up and at Umm-al-Tubal Townshend turned to give battle. He was seriously outnumbered and despatched the cavalry brigade to cover his right flank, which was in danger of being turned. Covered by its machine-gun section and S Battery, the Regiment deployed for dismounted action in the path of the advancing Turkish 51st Division. A firing line was formed and rapid fire was opened at 1,200 yards. The Regiment took a pride in its musketry and soon gaps were being torn in the opposing ranks. Still the Turks came on and S Battery, now engaging over open sights, would have been forced to withdraw to avoid being overrun had not the machine-gun section done considerable execution among its attackers; this was a particularly warm corner of the battlefield and within the space of a few minutes almost everyone in the section received some sort of wound – the section commander, Lieutenant

Pope, escaped lightly with a bullet through his helmet. This, however, was the turning point of the battle and shortly after the Turks broke off their attack, enabling Townshend to disengage and resume his withdrawal.

The column, close to exhaustion, moved slowly, and, as the day progressed, a force of Turkish infantry and cavalry, the latter estimated to number six squadrons, was observed closing in on its exposed flank. The Fourteenth were ordered to disperse them and the Regiment moved out at a steady trot. Eight hundred yards from the enemy, Hewitt gave the order to *Draw Swords* and swung his squadrons into line, immediately increasing the pace to a controlled gallop. The moral effect of this thundering wall of horses, topped with the rippling glitter of steel, was too much for the Turks; they wheeled about, dug in their spurs and bolted for safety. No further attempts were made to interfere with the column.

The battle of Umm-al-Tubal cost the Turks 1,500 casualties and Townshend's force less than one-third of that figure. The Regiment had five men killed and fifteen wounded, nine horses killed and twenty-five wounded.

Kut was reached on 2nd December and here the withdrawal ended. It is sometimes suggested that the exhaustion of Townshend's infantry made a protracted halt, and therefore seige, an inevitablity. Yet sleep and food can work wonders in a remarkably short space of time and those same 'exhausted' troops managed to dig six miles of trenches in very short order. Kut was not isolated until 7th December and Townshend could have continued his march downstream had he chosen to do so. Here, as the Turks themselves were to demonstrate, excellent defensive positions were available and reinforcements could have reached him. Since Townshend was an able commander, the reason for his illogical decision must rest within his own convoluted psyche. Having failed to capture Baghdad, he sought to compensate by re-creating the very conditions which had first brought him honour and reward. Indeed, no sooner had he entered the town than he sent a signal to Nixon; 'I mean to defend Kut as I did Chitral'. Nixon, the eternal optimist, endorsed his decision and promised that relief would arrive in two months at the most: unfortunately, Nixon lacked the means with which to make good his promise.

It was, however, decided that the 6th Cavalry Brigade would not remain in Kut and it marched out on 6th December. The Regiment was forced to leave its heavy baggage behind and with it a guard of sixteen men, who shared the fate of the doomed garrison. During the march downstream to Ali-al-Gharbi a skirmish took place with some of the enemy's irregular cavalry, who lost heavily.

The Turks attempted to storm Kut at Christmas but were so sharply repulsed that they decided to starve Townshend into surrender, simultaneously constructing positions that would thwart attempts to relieve the beleaguered fortress. In this they were fortunate that on the left bank of the Tigris the only approach lay along a narrow neck of land between the river and an extensive salt marsh, while on the right bank the ground was cut up by canals which in this, the wet season, proved a serious obstacle.

Every attempt to break through the Turkish lines failed, partly because of poor planning and partly because the enemy was fiercely motivated and determined that Townshend should not escape. The British and Indian infantry, on whom the burden of the fighting fell, displayed complete self-sacrifice and in the successive battles of Sheikh Saad, the Wadi Hannah and the Dujailah Redoubt incurred no less than 23,000 casualties; that this figure exceeded the total strength of the Kut garrison by almost 10,000 simply served to underline the criminal absurdity of the whole situation.

During these operations the cavalry covered the army's open flank, noting a marked reluctance on the part of their Turkish opposite numbers to come to grips. On one occasion the Regiment came within sight of Kut, but some of its most important tasks were the location of enemy trenches concealed in the flat landscape. This could only be done by drawing fire at close quarters; SSM Wright won a Distinguished Conduct Medal in this manner, galloping right up to a trench and then blazing away at its tenants with his revolver.

The siege of Kut was the first occasion on which attempts were made to supply an isolated garrison by air. Supply, however, came nowhere near meeting demand and, by April 1916, Townshend's troops were literally starving to death. The Turks were offered the sum of £1 million – later doubled – in exchange for a promise of parole, but they would have none of it; the most they would agree to was an exchange of hospital cases for an equivalent number of fit prisoners. On 29th April, Kut surrendered. The Turks kept their promise that Townshend himself would be treated as an 'honoured guest' but only half his men survived their harsh captivity.

So ended one of the worst disasters ever to befall British arms. It profoundly shocked public opinion and if it is little remembered today that is because it has been eclipsed by the even worse disaster of Singapore, with which it had much in common.

To Baghdad and Beyond

AFTER THE FALL of Kut there were no further major operations on the Mesopotamian front for a considerable period. Neither side was capable of taking the offensive; the British, because they were exhausted and sorely needed time to reorganise their administrative services; and the Turks, because Russian advances in Asia Minor and northern Persia compelled them to withdraw troops from the Tigris.

During these months, the Regiment took part in several aggressive demonstrations designed to limit such withdrawals and was also engaged in providing escorts for convoys and in keeping the ever-present Arab marauders at a distance. For a period the ravages of the climate meant that forty men per day were reporting sick with various illnesses and this seriously affected the numbers available for duty. As the year progressed, however, the general condition of the army improved beyond recognition, particularly after Lieutenant-General Frederick Maude assumed command.

Maude has been described as a cold, uncommunicative sort of man, but he was an excellent administrator and a sound strategist who enjoyed the confidence of the Imperial General Staff, and therefore received most of what he asked for. By December 1916, his army consisted of two corps and a cavalry division of two brigades, supported by a flotilla of modern gunboats and supplied by a fleet of specially designed shallow-draught steamers. The 14th Hussars and S Battery remained in the 6th Cavalry Brigade, but the 7th Lancers and 33rd Cavalry were relieved by two more Indian regiments, the 21st and 22nd Cavalry. Regimental firepower was vastly increased by the issue of one Hotchkiss light machine-gun to each troop, although the Maxim section was withdrawn and, with those of other regiments, formed into a brigade machine-gun squadron.

Maude's Turkish equivalent in Mesopotamia was Khalil Pasha, who had received Townshend's surrender. Because of this his reputation had soared, although the Kut operations had actually been conducted by a German officer, Field Marshal Kolmar von der Goltz, who had died of cholera a few days before the capitulation. For the moment, he was more concerned with events in Persia than with the Kut sector and was openly dismissive of warnings from Kiazim Karabekir, his local commander there, that Maude was daily growing stronger and was bound to attack sooner or later. Bluntly refusing Kiazim reinforcements, he told him that the Turkish positions below Kut were clearly impregnable and that the British, of whose military abilities he held a low opinion, would never be able to penetrate them.

For his part Maude knew that Kiazim possessed only three divisions, all of which were deployed well forward, and that behind these there were no reserves at all. His plan, therefore, was to destroy the Turkish army where it stood, using the indirect approach. The core of Kiazim's defences lay in the narrow corridor of land on the left bank of the river at Sannaiyat. On the right bank, however, the Turkish positions were not so strong and Maude decided to eliminate these one at a time and steadily work his way upstream past Kut and then cross the river. This would have the effect of trapping Kiazim unless he had the foresight to withdraw.

A sketch by Private Baggot showing the Regiment's first action in Mesopotamia.

Air raid on the Regiment's lines south of Kut. Baggot's drawings revealed acute powers of observation, a highly developed tactical sense and, as shown here, a wicked humour.

The British offensive began on 13th December with a feint which caused Kiazim to commit more troops to the Sannaiyat position, and then continued as Maude had intended. The step-by-step method worked and by the third week of February 1917 the right bank of the Tigris was in British hands for some miles above Kut. For Kiazim the *coup de grace* came when Sannaiyat itself, its defenders demoralised by an endless stream of shells from across the river, was stormed on 23rd February. His decision to abandon Kut and withdraw was merely bowing to the inevitable, but it came too late.

During this period, the Cavalry Division, commanded by Brigadier S. F. Crocker, not only secured the open British left flank but also took part in several operations which seriously unsettled the Turks. On 14th December, for example, a demonstration against the enemy's pontoon bridge on the Shumran Bend was so successful that the Turks hastily dismantled the bridge and towed it away, leaving themselves without communication across the Tigris for several days; it was a furthur interesting application of the indirect approach, in which the Regiment's B Squadron, commanded by Captain T. R. Bruce, galloped in open order to within 700 yards of the bridge, returning without casualties in spite of the heavy fire directed at them. On 24th December, the Cavalry Division marched south to capture and blow up a structure known as Gussab's Fort, a notable nest of Arab marauders. The same could be said of Hai Town, which the Division raided on 10th January. Here the Arabs seemed more intent on business than fighting, although some fell on the rear of the column as it withdrew, with sorry consequences for themselves. On 8th February, the Division carried out a dismounted diversionary attack at Shumran while the last fortified position opposite Kut, the Liquorice Factory, was stormed by the infantry.

Shumran had been selected by Maude as his own crossing point. On 23rd February, the 14th Division not only secured a bridgehead but also put in a pontoon bridge. The following day the Cavalry Division crossed and passed through the infantry to find Kiazim's troops streaming in retreat across the neck of the peninsula, covered by a strong flank guard. This withdrew during the night and on the morning of the 25th the Division became the advance guard of Maude's pursuit. During the day, however, both brigades became needlessly involved in a dismounted attack against the enemy's rearguard, partly as a result of conflicting orders from Maude, who had an unfortunate tendency to interfere in the details of his subordinates' operations.

Once again the Turks disengaged after dark and the pursuit was resumed next morning. By now three of Maude's gunboats, *Tarantula*, *Mantis* and *Moth*, were well upstream and at the Nahr Kellek Bend they caught up with Kiazim's column and engaged it with the sustained fire of their main armament and all their automatic weapons. This had the effect of turning the enemy's retreat into a rout. The 6th Cavalry Brigade closed in from the landward side but was unable to launch its own attack as it was immediately selected by the enthusiastic sailors as a fresh target, although S Battery was able to add to the Turks' discomfiture. At four o'clock the gunboats ceased firing and proceeded up-river, ending what for them had been a thoroughly satisfactory day by sinking two of Kiazim's three gunboats, *Pioneer* and *Basra*, and driving the third, *Firefly*, onto the bank. *Firefly* had been one of the vessels abandoned during Townshend's retreat but, having been towed off, she was able to continue the voyage to Baghdad she had started two years earlier, by coincidence under her original skipper.

Since the crossing of the Tigris, 4,000 prisoners had been taken and it was now clear that Kiazim's army was on the point of disintegration. 'The Turkish dead littered the road. There was every sign of panic and rout – guns and carts abounded, bullocks still alive and

B Squadron draw the enemy's fire at Sheikh Sa'ad, January 1916.

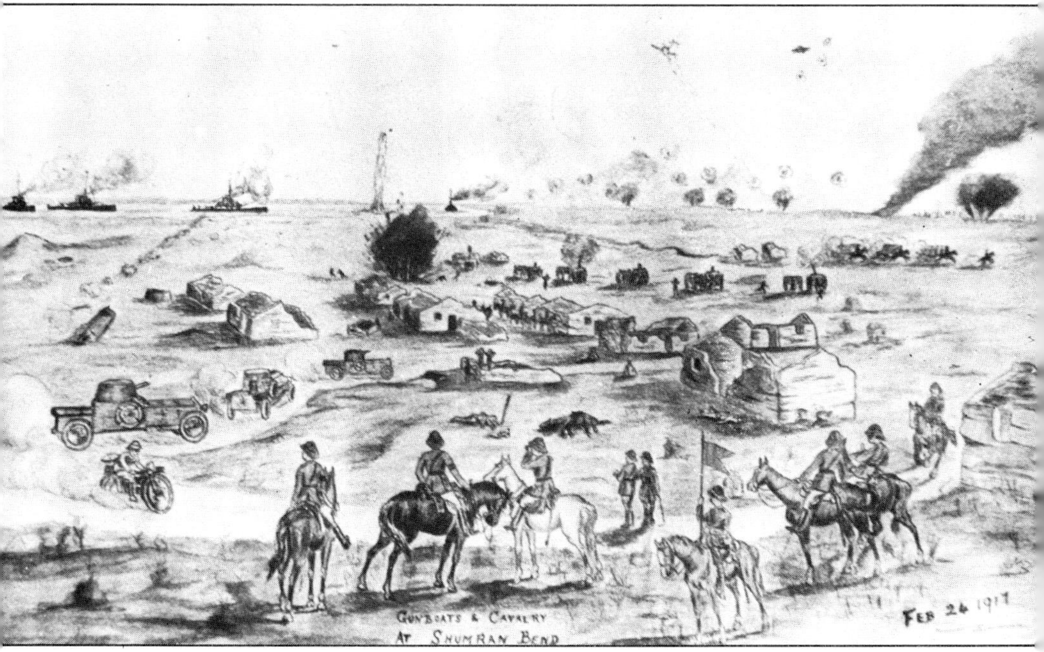

Baggot's detailed sketch of the advance north from Shumran. On the left are the gunboats *Tarantula*, *Mantis* and *Moth*, which are shelling the retreating Turks on the horizon while a Turkish gunboat flees up-river; in the middle distance S Battery RHA are going into action and Rolls-Royce armoured cars of a Light Armoured Motor Battery pass through a ruined village; RHQ is in the immediate foreground and a motor-cycle despatch rider is arriving with orders for Colonel Hewitt.

unyoked, broken wheels, cast-off equipment, overturned limbers, hundreds of live shells of various calibres scattered over the country for miles. Every bend of the road told of confusion and flight. Here a wrecked field post-office with Turkish money-orders and notes circling round in the wind; there office furniture, piles of books and documents; cartloads of small-arms ammunition, grenades, hats, boots, oil drums, things destroyed or half destroyed and dead animals. A brand new Mercedes motor-car abandoned for lack of petrol was an item of interest to the weary pursuers. White columns of smoke ahead, telling of further destruction, and enough litter by the road to keep an army in fuel for weeks. A whole battery of 12-cm field-guns was found half buried, their breech blocks removed.'*

Aziziyeh was reached on 1st March and here a halt was called while the supply line was re-established. The advance was resumed on 5th March and at Lujj the 7th Cavalry Brigade (13th Hussars, 13th and 14th Lancers and V Battery RHA) struck a freshly entrenched enemy position. The 13th Hussars attacked without hesitation, killed many of the Turks and captured others, but at a cost of ninety casualties. Coming up on the right, the 6th Cavalry Brigade committed the 14th Hussars to a dismounted attack against the enemy's flank. Believing that they were on the point of being encircled, the Turks pulled out during the night, having informed their superiors that they had been assailed by two cavalry regiments and an entire infantry division. The Lujj position was a naturally strong one and needed only a few more hours' work before it would have been capable of delaying Maude for several days.

On the 6th, the Cavalry Division crossed the deserted battlefield of Ctesiphon and by the following evening had reached the Diyala, a tributary of the Tigris which flowed in from the north-east; from here the domes and minarets of Baghdad were clearly visible.

To capture the city, Maude decided to employ the same indirect approach he had used against Kut. While the 13th Division crossed the Diyala, the Cavalry Division and two infantry brigades re-crossed the Tigris and began to move up its right bank. Khalil Pasha was now conducting the defence in person and although his troops were offering determined resistance he was seriously unsettled by the cavalry's threat to his rear. In the nick of time, he managed to squeeze out of the trap. The 14th Hussars reached Khazimain, four miles north of Baghdad, during the morning of 11th March, to find the enemy gone, and Colonel Hewitt accepted the surrender of its notable citizens. Simultaneously Maude rode into Baghdad, receiving the exaggerated plaudits of those same Baghdadis who, less than a year earlier, had kicked, pelted and spat upon Townshend's skeletal survivors as they had staggered through.

For the next six weeks, Maude concentrated on pushing the beaten Turks away from Baghdad to the north and east. However, fresh Turkish troops were arriving at the front and in particular Ali Ishan Pasha's XIII Corps from Persia was moving down the Diyala to menace the British right flank. The Cavalry Division was moved into a blocking position at Delli Abbas where, on 27th March, the weight of the enemy's attack fell on the 6th Cavalry Brigade's position between the Diyala and the Al Khalis Canal.

During the action, the 22nd Cavalry were forced to retire through the Fourteenth's B and C Squadrons, who in turn were compelled to fall back into line with the 21st Cavalry's two squadrons, behind whom was deployed the brigade's stop-line, consisting of 22nd Cavalry, the Fourteenth's A and D Squadrons and the brigade machine-gun squadron. It was a textbook example of firepower and mobility combining to erode the strength of an

---

**Historical Record of the 14th Hussars,* Vol 2, p361.

enemy's attack. Troops fought dismounted until ordered to withdraw, then the led-horses were brought up and they galloped back to the next bound. One troop of C Squadron had an unpleasant moment when forced to remain in position longer than it cared, to cover the withdrawal of two of 22nd Cavalry's ambulance waggons, but casualties were remarkably light. The Turks, however, suffered very severely and by evening had been fought to a standstill.

An almost identical battle took place on the opposite bank of the Al Khalis Canal on 11th April. On this occasion, however, the 21st and 22nd Cavalry had been detached to other duties and 6th Cavalry Brigade consisted only of the 14th Hussars, S Battery and the machine-gun squadron. Similar tactics were employed, the enemy's main body being engaged at 1,300 yards and his skirmishers at 700–900 yards, the Regiment retiring by bounds past the right flank of 7th Cavalry Brigade, which blocked the Turks' further advance until infantry of the 13th Division arrived to establish a firm front.

Ali Ishan made one further half-hearted attempt to break through on the Diyala sector and then withdrew through the Jebel Hamrin at the end of the month. This coincided with Maud's capture of Samarra on the Tigris and brought major operations in Mesopotamia to a close for several months.

A wounded officer is brought in during the action at Delli Abbas. Note the troop firing line on the crest to the left, with the led horses in dead ground behind.

Officer and troopers, 14th King's Hussars, Mesopotamia, 1917.

Ramadi

DURING THE HOT SEASON, the Cavalry Division established a camp at Chaldari, seven miles north of Baghdad. In July, temperatures soared on several occasions to 122 degrees in the shade, dropping towards the end of the month to a mere 108 degrees, which one officer described as being 'quite chilly!' It was unfortunate that this period was chosen for a raid by 7th Infantry Brigade against the 1,000-strong enemy garrison at Ramadi on the Euphrates, for, although the troops travelled by night and in vans, the operation failed because of a combination of dust storms, heat exhaustion and stubborn Turkish resistance. Because of this, the Turks wasted no time in reinforcing their garrison to a strength of approximately 4,000.

With the arrival of cooler weather, Maude decided to eliminate this threat to his flank and, in September, the 15th Division, the 50th Brigade Group and the 6th Cavalry Brigade were detailed for the task. The plan was for the enemy's main position at Ramadi to be attacked by the infantry while the cavalry cut the Turks' line of retreat along the Aleppo road. The infantry attack went in during the afternoon of 28th September and its success convinced the Turkish commander, Ahmed Bey, that he must break out.

Nightfall found the 6th Cavalry Brigade deployed in a crescent which both blocked and covered the Aleppo road. The brigade commander, Brigadier P. Holland-Pryor, had placed three of the Regiment's squadrons, D, B and A, on the left or river flank, with two squadrons of the 21st Cavalry on their immediate right; in the angle of the position was the brigade machine-gun squadron with three squadrons of 22nd Cavalry on their right, lining a low ridge which protected the led-horses; then came brigade headquarters with V Battery RHA, the extreme right wing being covered by C Squadron 14th Hussars and a 22nd Cavalry squadron; in immediate reserve and deployed some way behind the left flank were two squadrons of 21st Cavalry. Altogether, including horse-holders and those employed on other duties, the Regiment's strength at Ramadi amounted to 170, of whom 120 were in the firing line.

Colonel Hewitt had established a standing patrol under Lieutenant G. G. Maule on the road into Ramadi. During the early hours of 29th September this detected the first signs of the Turkish withdrawal and quietly walked in to alert the squadrons. At 2.45 am Hewitt gave the order to fire and when this was taken up all along the line the Turks realised that they had walked onto a killing ground. Nonetheless, they continued to press their attack with the utmost determination and some of their dead were found only fifty yards from the firing line. The cavalry's initial advantage declined when one of the enemy's barges on the river was set ablaze, the glare degrading night vision without being sufficiently intense to provide illumination. When the moon began to set behind the brigade's position it threw men and animals into silhouette and in consequence casualties began to mount. After moon-set, however, firing died away and dawn revealed that the Turks had retired into Ramadi. They had sustained over 600 killed and wounded in the action and at 11.00 am Ahmed Bey surrendered; 3,545 prisoners were taken, as well as thirteen guns, heavy automatic weapons, ammunition, transport vehicles and barges.

The Regiment's casualties were twenty-seven killed and wounded. Among these was Colonel Hewitt, who was mortally wounded at the height of the battle; universally liked and respected, his loss was deeply felt. 'It was difficult,' wrote one officer of the march back to Baghdad, 'to think of anything but the empty place in front of the Regiment, where Colonel Hewitt's charger, saddled and bridled, was being led by his faithful orderly.'

In recognition of the 6th Cavalry Brigade's achievement, General Maude paid a personal visit to Chaldari Camp on 8th October, commenting in his speech, 'I do not think that it is too much to say that the fighting at Ramadi was an instance of as clean and businesslike a military operation as one could wish to see.' The anniversary of the battle is celebrated by the Regiment every year in the Corporals' Mess.

The Cavalry Division spent the remainder of the year engaged in minor actions, reconnaissance and patrolling. On 6th November, the Turks were defeated at Tekrit and forced to retire even further up the Tigris; during this action the cavalry manoeuvred against the enemy's flank and rear, but the lessons of Ramadi had clearly been learned and the Turks had taken considerable trouble to see that their line of retreat remained open. This, sadly, was Maude's last victory, for on 18th November he succumbed to cholera and Lt-General Sir W. R. Marshall assumed command of the army.

Lt-Colonel R. W. Hewitt, DSO,
who commanded the 14th King's Hussars in Mesopotamia.
Colonel Hewitt died from wounds received at Ramadi.

Bridges' Column

IN JANUARY 1918, Lieutenant-Colonel E. J. Bridges, commanding 14th Hussars, was advised that the Regiment was to be detached from the 6th Cavalry Brigade to form the major element of a column which would operate for a considerable period beyond the Persian frontier; also included in the column were one section each from S Battery and the machine-gun squadron, a pack wireless section and a supply and transport section. The column remained in Persia for most of the year, squadrons being separated by hundreds of miles of mountainous, trackless country from RHQ and from each other. The confused nature of their operations was such that even twelve years after the event Colonel Bridges was forced to admit that 'there is considerable difficulty in making the story either comprehensible or connected. The military and political reasons which necessitated the expedition were only vaguely understood at the time, nor were they clearly explained subsequently'.*

At its simplest, the overall situation can be summarised as follows. Because of the oil factor, a pre-war agreement between Russia and Great Britain divided Persia into respective northern and southern spheres of influence. During the early months of the war, the Russians had occupied northern Persia and the Turks had mounted a counter-invasion. Because of the 1917 Revolution, the Russian armies in Persia and Asia Minor had disintegrated and the Turks, though defeated in Mesopotamia and Palestine, had begun to advance into Russia, making the rich Baku oilfield their primary objective. For such a prize to fall into the hands of the Central Powers would have been the equivalent of a major defeat in the field and it was decided that the Turks must be stopped at all costs. This was no easy matter, for it will be recalled that Baku lay on the opposite shore of the Caspian Sea.

The officer chosen for this apparently impossible task was Major-General Lionel Dunsterville under whom such troops as Marshall could spare were grouped together as Dunsterforce. An old school friend of Kipling, Dunsterville was actually the prototype of the resourceful Stalky of *Stalky & Co.* Incredibly, he succeeded in getting a British garrison into Baku and in holding it long enough to deny the enemy its oil before the war ended. In this context the function of Bridges' Column was first to assist in opening Dunsterville's axis of advance from Hamadan to the port of Enzeli, and then prevent interference by Turkish-inspired tribes with his long line of communication. There were, however, many other factors for Bridges to consider in addition to the purely military aspects of the affair, for, in this part of Persia, society had degenerated into an armed maelstrom in which political, racial and religious differences were being openly settled by violence, the whole messy pot being stirred at will by Turkish, German and Bolshevik agents. To further confuse the issue, people changed sides with bewildering frequency. For example, in March, B Squadron carried out a successful punitive expedition against the Kalhurs, but when, in April, A Squadron mounted a similar raid against the Sinjabis, the Kalhurs were only too pleased to join in – not because of any newly-discovered loyalty, but simply because they disliked their neighbours. Add to all this a famine that was sweeping the

*Historical Record of the 14th Hussars, Vol 2, p432.

country and it will be readily understood that squadrons needed all their wits about them to survive. Several large, and numerous small, actions were fought but here it is only possible to follow the mainstream of events.

In April, D Squadron, under Captain Featherstonhaugh, took part in an attack on the Turkish post of Kifri, taking seventy prisoners. Meanwhile, C Squadron, commanded by Captain A. V. Pope, moved to Hamadan and then on to Kasvin. Here they came under the orders of a Russian general named Bicherakov, who was anxious to return home with his Cossack infantry and cavalry and join in the civil war on the side of the Whites. Unfortunately, Bolshevik intrigue had seen to it that the road between Kasvin and Euzeli was blocked at Menjil Bridge by a force of Jangalis commanded by a Persian nationalist named Kuchik Khan, who had a German military adviser, and it was clear that a battle would have to be fought.

At this point two twin-turreted Austin armoured cars were placed temporarily under Pope's command. They belonged to an extremely experienced unit known as Duncars, which was actually part of the Machine-Gun Corps, but which for the past two years had served inside Russia as a unit of the Royal Naval Armoured Car Division under the command of a swashbuckling Member of Parliament named Oliver Locker Lampson. Given that in Persia the unusual could safely be regarded as the normal, the Hussars would have been only mildly surprised to learn that the lowest rank in Duncars was Sergeant, and prior to that had been Petty Officer.*

The attack on Menjil Bridge took place on 16th June. Two aircraft strafed the enemy trenches, which then came under fire from the Russian artillery. The Cossack infantry attacked in the centre while the cavalry charged the enemy's right flank and the armoured cars tore straight down the road at the bridge, guns blazing. The Jangalis fled, losing heavily, and forty prisoners were taken. A curious discrepancy arises concerning C Squadron's part in the battle: the full Regimental history states that it remained in general reserve, yet independent sources suggest that it was involved in the general attack and pursuit – given the weight of evidence, the latter seems not improbable.

C Squadron remained at Menjil for two months and then received orders to proceed to Zinjan. While *en route* it was ambushed by Kuchik Khan's men in a defile near Kaskal. The attack was beaten off by the fire of the Hotchkiss guns and the squadron emerged unscathed, having taken three prisoners and a quantity of ammunition.

Dunsterville was now installed in Baku and the Turkish 11th Division marched south from Tabriz with a view to severing his communications in north Persia. Very few troops were available to counter this move but these were contesting the enemy's advance beyond Mianeh and C Squadron was ordered forward to join them, together with two artillery batteries, a company of the Hampshire Regiment and two armoured cars from a Light Armoured Motor Battery.† The squadron's first contact with the enemy occurred on 6th September and for the next week it fought a more or less continuous rearguard action. On one occasion it took the Turks seven hours to advance a mere two-and-a-half miles against it and they were finally brought to a standstill at the foot of the Kuflan Kuh Pass. Though at times deadly tired, and with their clothing and boots worn to tatters, the only serious com-

---

*The history of this very remarkable unit can be found in the author's book *The Czar's British Squadron*, published by William Kimber.

†The LAMBs were noted for their dash, efficiency and independent outlook, their usual equipment being the Rolls-Royce armoured car. These particular cars probably belonged to the 6th LAMB and C Squadron carried out a number of successful joint operations with them.

plaint made by Pope's troopers during this period related to an issue of what they considered to be obnoxious 'fish jam'; this, of course, was caviare, for which the area is famous. In October, the squadron marched to Kasvin, where it was joined by B Squadron.

Throughout the summer, the rest of the Regiment had also been actively employed, although not to quite the same degree. A Squadron, based at Senneh, had established a firm grip on the surrounding country. B and D Squadrons, operating from Bijar, were mainly concerned with covering the flight of the Christian Jelus, whose homeland lay in the region of Lake Urmia to the north, against whom the Turks and their Kurdish allies were waging a murderous war of extermination. While so engaged, Sergeant Paddy Hallard of D Squadron won a well deserved Distinguished Conduct Medal when he led his twelve-man patrol to the rescue of a small party of troops who had lost their officer, shot off most of their ammunition, and were in danger of being cut off. Hallard, outnumbered by twenty-five to one, covered the party out and fought a series of stubborn rearguard actions until the rest of the squadron arrived.

By October, Turkey was reeling under the hammer blows of Allenby's victory at Megiddo in Palestine and the renewed British offensive on the Tigris, and on the 30th she capitulated. Bridges' Column commenced its return journey to Mesopotamia in December, accompanied by Armenian irregulars, Russian sailors, a Cossack prince, refugees and an Indian pilgrim – 'a rare circus', as somebody said.

Amalgamation

THE REGIMENT arrived home in 1919 and the following year joined the British Army of the Rhine at Cologne. Because of friction along the German–Polish border, A Squadron was detached for peace-keeping duties in Silesia and remained there until June 1922. At this period raging inflation destroyed the German economy so thoroughly that by the end of the Regiment's three-year tour the rate of exchange had reached 350 billion marks to the pound.

Meanwhile, the defence economists' axes were swinging in every possible direction. The experience of the Western Front had shown that cavalry was unlikely to play a major role on a future European battlefield and a reduction in the number of Regular cavalry regiments from thirty to twenty-two was ordered. This was achieved by a series of mergers, and, on 1st October 1922, the 14th King's Hussars amalgamated with the 20th Hussars.

The 20th Hussars

Orlando Norrie's painting of the 20th Hussars in line, advancing in review order.

Distinguished Forebears

BEFORE JOINING the Army's permanent establishment, the Regiment had three previous incarnations. The 20th (Inniskilling) Light Dragoons were raised during the Seven Years' War and served in Ireland until 1763, when they were disbanded. The Regiment was raised again as 20th Light Dragoons in 1779 at Bury St Edmunds and served in England throughout the American War of Independence, during which Great Britain found herself not only at war with her former colonies, but with most of Europe as well. When peace was concluded in 1783 it was again disbanded but was reactivated nine years later for service in the West Indies on the outbreak of the slave rebellion. On this occasion it was raised by Lieut-Colonel A. Farington Gardiner and brought up to strength by drafts from other units, including a contingent from the Fourteenth. Known originally as the 20th (Jamaica) Light Dragoons, the Regiment remained in the West Indies until this difficult campaign had been brought to a satisfactory conclusion. However, once the threat posed by the slaves had been removed, the authorities in Jamaica, who were responsible for the financial upkeep of the Regiment, requested that it be withdrawn, and it returned home in 1800. At this period the Regiment wore the standard light dragoon uniform faced, like that of the Fourteenth, with orange, although the lace adornments were gold rather than silver.

In 1805, the Regiment embarked for foreign service again, two squadrons taking part in an expedition to the Cape of Good Hope, while three were sent to the Mediterranean theatre of war. The object of the former was to sieze Cape Town from the Dutch, who were allied with Napoleon, and an interesting account of this and its sequel was written by Sergeant Landsheit of the Twentieth, who is best described as something of an old soldier – in both senses of the word.

The expedition reached Table Bay on 4th January 1806, but such fighting as there was had ended in a Dutch capitulation before the Regiment was involved. Having accomplished their primary mission the expedition's commanders, Admiral Sir Home Popham and General Sir David Baird, immediately sought further means to employ the forces at their disposal. Both had a taste for prize money and, without bothering to inform the home government, they decided to attack the Spanish colonies in South America, Spain being, for the moment, another ally of Napoleon's. In June, an advance party, consisting of the 71st Regiment and four guns, accompanied by an officer and six men of the 20th Light Dragoons, siezed Buenos Aires by *coup de main* and looted its treasury, making Baird the richer by £30,000, Popham by £6,000 and each private soldier by £18.6.0d. It did not take the enraged Argentines long to discover that they had been forced into flight by a comparative handful and they soon rallied in large numbers, surrounded the little force and ultimately compelled its surrender.

Baird arrived in the River Plate with his main body in October and landed near Maldanado, in present-day Uruguay. A battle was fought near the town, the local troops taking to their heels after the first vollies; during this engagement, which ended with the 21st Light Dragoons chasing the enemy through the streets, the Regiment fought dismounted. No doubt inflamed by stories of the loot which could be acquired in South

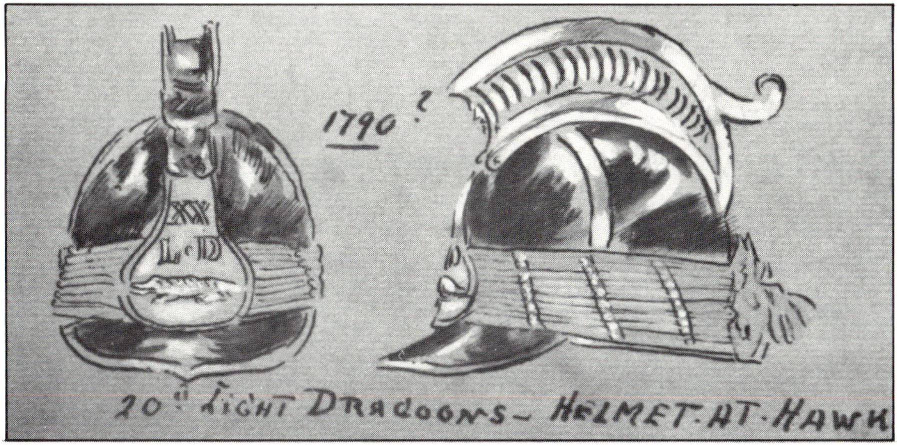

Sketch of tin tropical helmet worn by the 20th (Jamaica) Light Dragoons; this would normally have been ornamented with a trailing plume of red horsehair. The alligator badge remained in use until the Regiment returned home in 1800.

20th Light Dragoons Regimental Medal. The wreath consists of pineapples.

America, the British soon got completely out of hand and subjected Maldanado to three hours of drunken pillage. During this unfortunate episode Sergeant Landsheit stood a little apart, making disgusted noises but doing little to influence events. Once order had been restored, however, he quickly assembled a collection of elegant female underwear which he then distributed among the Regiment's womenfolk, hinting smugly at the favours received in return.

The government was far from pleased with the high-handed way in which the whole business had been initiated and Popham was court-martialled. Nonetheless, the Argentine gold was very welcome and it was decided to continue with operations in South America. In January 1807, General Sir Samuel Auchmuty arrived at Maldanado with reinforcements and assumed command. He advanced up-river on Monte Video where, on the 19th, he found the enemy drawn up on a ridge outside the city. An immediate attack was ordered by the cavalry and during the subsequent charge the Twentieth's two squadrons captured three guns and took part in the general pursuit of the Spaniards. The city was then invested, a sortie by its garrison being repulsed with heavy loss; according to Landsheit, the enemy had fortified their courage with liquor, which helped them not at all. Monte Video was stormed on 3rd February at a cost of 100 British lives, the Spanish loss being eight times that figure.

Later in the year, a second attack was made on Buenos Aires, but this failed because of the incompetence of the British commander, General Whitelock. The Twentieth were not involved and returned home in September, one squadron being shipwrecked off the Needles.*

Meanwhile, the Regiment's other wing had landed in Sicily, then a part of the Kingdom of Naples, with which Great Britain was allied. A detachment of sixteen men took part in an expedition to the mainland of Italy and were the only British cavalrymen present at the battle of Maida where, on 4th July 1806, Sir John Stuart utterly routed a French army commanded by General Reynier.

In 1807, the three squadrons formed part of a small force which was despatched to Egypt under Major-General M'Kenzie Fraser to safeguard British interests, as the country's effective ruler, Mehemet Ali Pasha, was openly sympathetic to Napoleon. Alexandria was occupied without undue difficulty, but Fraser felt that he must also secure the town of Rosetta to consolidate his position. Rosetta, however, resisted all his efforts to take it and fresh Turkish troops marched to its relief. A violent clash took place at El Hamid on 19th April in which Captain Delaney's squadron charged a far larger body of the enemy's cavalry. Sheer weight of numbers brought the charge to a standstill and Delaney and twelve others were captured. The remainder of the squadron dismounted and formed square, holding off the Turks with the fire of their carbines. The situation was desperate and it was clear that sooner or later the square would be wiped out. Private Tremble volunteered to ride for help, although few would have imagined that he stood the slightest chance of success. Nonetheless, Tremble quite literally cut his way through and returned with two infantry companies, whose fire drove off the Turks.

After El Hamid, Fraser retired to Alexandria and, after an accommodation had been reached with Mehemet Ali, the expedition returned to Sicily, from whence all but four of the Twentieth's troops sailed home.

---

*During the voyage south a transport had been wrecked on the coast of Brazil, one of the very few casualties being an artillery Brigadier who had ballasted his pockets so well with doubloons that he went straight to the bottom. 'Misguided and fatal avarice,' mused Sgt Landsheit, who knew best about many things.

Details of badge and uniform worn by the Regiment
in the Peninsula and elsewhere.

A composite view of the Battle of Vimiera. Colonel Taylor is shown leading out the Twentieth in the bottom right-hand corner. The Regiment's charge would take it diagonally across the picture to the enclosure at the top left.

National Army Museum

The following year the combined Regiment, commanded by Lieutenant-Colonel Charles Taylor, embarked with Wellington's army for Portugal.* After landing at Mondego Bay they joined Brigadier Fane's light infantry brigade, which formed the advance guard of the British advance on Lisbon. A small French force was defeated at Rolica, where the ground was so rocky and broken that the Twentieth were not directly involved in the fighting, and Wellington resumed his march, halting at Vimiera.

It was now known that Marshal Junot, the French commander in Portugal, had concentrated his troops and was marching on Vimiera along the road from Torres Vedras. The Regiment sent out mounted patrols and it was one of these, commanded by Sergeant Landsheit, which first spotted the French on the evening of 20th August. Whatever reservations one might have regarding Landsheit, he was good at his job and galloped in to report to Fane, who sent him straight on to Wellington's headquarters, which he reached about midnight. Here he found Wellington and his staff in their mess, sitting back to back on a long table, presumably because of a shortage of chairs, and in a surprisingly relaxed mood. He delivered his report in a smart and soldierly manner, upon which Wellington complimented him, and was then sent along to the kitchen for something to eat.

Wellington had the army roused quietly and took up the positions from which he repulsed Junot's attacks the following day. The French came on in columns and were met by the British in line, a formation which made best use of the available firepower. While the battle raged, Colonel Taylor repeatedly asked Fane for permission to counter-attack with the Twentieth, but was refused on the grounds that the time was not yet ripe; indeed, with the exception of two troops of Portuguese mounted policemen, the Regiment was the only cavalry unit Wellington possessed and could not be committed lightly. In due course, however, the defeat of a particularly strong enemy attack convinced the Brigadier that the moment had come.

'*Now* we want you, 20th!' he shouted. 'Forward and charge and show them what you are made of!'

The Regiment received a tremendous cheer as it trotted out through the British line. Increasing its pace steadily it thundered over the retreating French infantry as though they were of no account, and the impression given is that it regarded its real target as being several dragoon and chasseur regiments which had moved forward to cover the latter's withdrawal. The French cavalry had no time in which to work up enough speed for a counter-charge and the Twentieth slammed into them at full gallop, 'cutting and hacking and upsetting men and horses in the most extraordinary manner possible, till they broke and fled in every direction.'

British cavalry at this period were frequently criticised for their lack of tactical control after a success, and the Twentieth's charge at Vimiera is sometimes quoted as an example of this. It may indeed have been in Colonel Taylor's mind to have the *Recall* sounded, but he was shot through the heart just as the French broke, and the Regiment maintained its pursuit of the beaten fugitives. Some grenadiers lining a fence were cut down to a man as a torrent of light dragoons swept over the obstacle. Unfortunately, this brought them into an enclosed field surrounded by enemy infantry, behind whom the French cavalry rallied and returned to the attack. A furious melée took place in and around the field. During this Corporal Marshall, who had already established his reputation as a swordsman in South America, found himself engaged simultaneously with four dragoons. Marshall was riding a

*At this stage of his career Wellington was simply Lieutenant-General Sir Arthur Wellesley.

rowdy stallion which lunged, kicked and bit so effectively that it prevented the Frenchmen from concerting their attack. At length the Corporal felled an opponent on his offside with Cut One; he was a strong man, but even so he must have stood in his stirrups to deliver this near vertical stroke, for such was its power that it clove through the dragoon's brass helmet and into his skull as far as the palate. Wheeling to the left, Marshall eliminated one of his enemies on the nearside with the horizontal backhand Cut Six to the face, which sent the man reeling over his crupper. Then, suddenly, the two survivors had gone, and with them the rest of the French, for Brigadier Fane had sent the 50th Regiment doubling forward to the relief of the hard-pressed Twentieth, whose charge had carried them almost to the position occupied by Junot himself.

The battle of Vimiera, which cost the Regiment twenty killed and twenty-four wounded, was such a disastrous defeat for the French that Junot agreed to evacuate Portugal, to the fury of Napoleon, who began pouring troops into the Iberian peninsula. Wellington, too, was reinforced and the Twentieth looked forward to adding to their laurels in what promised to be a long war. During the operations on the Duoro in 1809 one squadron each of the 20th and 16th Light Dragoons, under the command of Major Blake of the Twentieth, had a sanguinary clash with the French rearguard near Grijo, killing a large number and taking 100 prisoners, British casualties being twenty-one killed and sixty-nine wounded, about half of those involved. Shortly after this, the Regiment received the unwelcome news that it would be rejoining its comrades in Sicily, which had now become something of a military backwater.

The Twentieth remained in Sicily for three years, detachments taking part in the capture of various French-held islands in the Tyrrhenian and Ionian Seas, but was otherwise engaged in garrison duties. In 1810, one of the Regiment's patrols gave early warning of an attempt by the French to cross the Straits of Messina, the result being that this failed with the loss of over 1,000 prisoners.

In 1812, Wellington asked the commander in Sicily, Lord William Bentinck, to effect a landing on the north-eastern coast of Spain, the object being to tie down General Suchet's French army and so prevent it interfering in operations elsewhere. The Twentieth formed part of this expedition, which landed at Alicante in August.

Strategically, the venture was a success, despite the chaos caused by the nervous disposition of its commander, Lieutenant-General Sir John Murray, whose orders and counter-orders probably confused Suchet as much as they did his own troops. Bentinck himself assumed command in June 1813, and thereafter operations were conducted on more normal lines. The war tended to be one of manoeuvre, since neither army was strong enough to force a decision on the other, yet could not afford to be defeated in the field; there were, however, a number of small but violent clashes.

The Regiment was brigaded with a regiment of Brunswick Hussars and a unit simply referred to as the Foreign Hussars, which had been recruited in Sicily and was commanded by Captain Jacks of the Twentieth with Sergeant Landsheit as his sergeant-major. In total, the brigade numbered approximately 400 sabres and was commanded by Lord Frederick Bentinck, Lord William's brother.

Its most serious engagement took place at Villa Franca on 13th September 1813. On this occasion Suchet felt strong enough to take the offensive and would have broken through the British rearguard had not the Twentieth and the Brunswick Hussars charged the French cavalry, whose strength was estimated at 2,000. In the desperate fight against odds which followed, the Twentieth displayed the same ferocity it had shown at Vimiera and it

Guidon carried by the Regiment during the Napoleonic Wars.

was the French who turned from the melée, having sustained 300 casualties, including their brigade commander, who had been wounded and unhorsed in single combat with Lord Frederick Bentinck. Nonetheless, the Regiment had not escaped lightly, its losses being nine killed, twenty-seven wounded and six missing, while the Brunswick Hussars suffered a total of fifty-three casualties. When the French cavalry returned to the attack they were easily driven off by the steady fire of the infantry, and Suchet broke contact.

Wellington was now pushing through the Pyrenees and by early 1814 the French had abandoned Spain altogether. In Italy, however, they retained various possessions, contrary to the spirit of agreements which were signed on Napoleon's first abdication, and Bentinck was determined to eject them from these. In March 1815, four troops of the Regiment took part in an expedition to capture Genoa where, after some desultory fighting in the approaches to the city, the French garrison agreed to march out with the honours of war.

After this the Regiment concentrated at Palermo for the voyage home, having earned the battle honours Vimiera and Peninsula. It was reduced to peacetime establishment and was serving in Ireland when, as a result of inevitable post-war economies, it was disbanded in December 1818, although a thread of continuity with its subsequent history was provided by a large number of its members who enlisted in the service of the East India Company.

Nobody's Own

AFTER THE GREAT MUTINY direct responsibility for the government of India was assumed by the Crown, whose representatives took it for granted that John Company's European regiments would accept transfer to the Queen's service without demur. In fact the Europeans, jealous of their privileged rates of pay and conditions of service, objected very strongly to the idea and staged the series of violent demonstrations which became known as the Little Mutiny. Fortunately, commonsense prevailed and reasonable concessions were offered to those willing to transfer, while those who were not were provided with a discharge and a free passage home; ironically, many of the latter enlisted anyway, being unable to settle in civilian life.

Thus, after an absence of forty-three years from the Army List, the 20th Light Dragoons were reconstituted on 2nd May 1861 with volunteers from the 2nd Bengal European Light Cavalry, and later brought up to strength with drafts of men who preferred to remain in India rather than return home with their own regiments, including a substantial number of veterans from the 14th Light Dragoons.

Some months later the Twentieth were advised that they had been re-designated Hussars and that their principal uniform distinctions would be a crimson busby bag and yellow plume. Initially, of course, hussar uniform was not available for issue and the Company's silver-grey light cavalry uniform was worn in its place. An interesting contemporary photograph shows a group of the Regiment's officers, many of them sporting fashionable

Officers of the 20th Hussars photographed at Campbellpore in 1866. At this period the Regiment still retained a hint of its John Company origins.

mutton-chop whiskers. Granted that having one's picture taken was then viewed as being a most serious matter, these men regard the camera with a hard, speculative stare which tends to emphasise that differences did exist between the Queen's and Company officers. In time such differences would disappear, although the Regiment always maintained a jaunty independence which is reflected in its self-conferred nickname – Nobody's Own. To the rest of the Army the Twentieth was known as the Xs, because of its Roman numeral badge.

The Regiment was mobilised for service on the North West Frontier during the Ambeyla campaign of 1863 and again during the Hazara campaign of 1868, but on both occasions it was employed on line-of-communication duties. It embarked for England in 1872 and two years later was officially recognised as the heir to the old Twentieth which had served throughout the Napoleonic Wars. This immediately entitled it to the Peninsula battle honour, although for some reason it had to wait until 1890 before it was credited with Vimiera as well.

The Sudan 1884–1889

'WHEN ALLAH MADE THE SUDAN,' goes the bitter Arab proverb, 'He laughed.' This cruel land was actually a province of Egypt but had been so corruptly administered that a religious fanatic named Mohammed Ahmed Ibn Al-Sayed Abdullah found no difficulty in having himself proclaimed Mahdi (Messiah) and in waging an extremely successful Holy War against the hated Turks, as the Egyptians were known. Every attempt to put down the rising ended in the massacre of the troops involved and, since at that period Egypt itself was virtually a protectorate of the United Kingdom, some British involvement became inevitable. General Charles Gordon was sent to the Sudan to co-ordinate the withdrawal of the surviving Egyptian garrisons, but in 1884 he decided to disobey his orders and hold Khartoum against the Mahdi's dervish army. Gordon was a popular figure at home and, much against its will, the government was forced to mount a relief expedition to rescue him. This included a Camel Corps, for services in which two officers and forty-three men of the Twentieth volunteered. The expedition fought its way through to Khartoum in January 1885, only to find that the city had fallen days earlier and that Gordon was dead.

The dervishes were formidable opponents who were possessed of religious fervour and did not care whether they lived or died. Particularly formidable were the Beja hill tribes, who were better known as Fuzzy Wuzzies because of their high-crowned hairstyle. Kipling immortalised the Fuzzy Wuzzy in his poem of the same name, describing him as

'An injia-rubber idiot on the spree,'

but expressing sincere admiration for the one man on Earth who had broken a British square. As a young journalist in India, Kipling spent much of his spare time in the company of soldiers, and he knew how they thought and talked. They had fought Afghans, Zulus and Boers, but never anyone like Fuzzy Wuzzy, who ignored the disciplined vollies that had dropped lesser men in their tracks, and emerged from the smoke yelling and waving his fearful two-handed sword around his head. Nor was he the least frightened of cavalry. One of his weapons was a heavy, weighted stick which he would throw at a horse's legs, bringing it down, the stunned rider being promptly despatched with sword or spear. Possessed of tremendous agility, he would bob around a horse's head 'like a rubber ball', dodging sabre cuts but being dangerously quick with his return stroke. If ridden down he would remain on the ground, beyond reach of the curved sabre, and hamstring or stab upwards. The experience contributed much to the adoption by the cavalry of the straight sword and also demonstrated that against such opponents the lance was a better weapon at close quarters.

So startling were the dervish successes that an invasion of Egypt seemed a distinct possibility. The British government therefore contributed troops to the defence of the Nile frontier at Wadi Halfa and also provided a garrison for the port of Suakin, which lay on the Red Sea inside the Sudan. The latter included two squadrons of the Twentieth, which arrived in March 1885. It had been hoped to link Suakin with Berber on the Nile by means of a strategic railway. Unfortunately the hinterland was controlled by Hadendowa Beja under their Emir, Osman Digna, otherwise known as The Ugly, whose power would have to be broken before the work could proceed. With this object in mind the garrison, led by its cavalry, marched in square to Hasheen where a series of ferocious engagements was fought, the results of which were inconclusive, although the dervishes suffered by far the heavier loss. Osman, a devout man who had once threatened British troops with extermination unless they instantly embraced Islam, did not take part in the fighting, but contributed a number of prayers from a suitable distance. It was, however, now clear that the Berber railway was not a practical proposition and the garrison retired to Suakin which, for the time being, Osman left severely alone.*

Following Hasheen, the two squadrons returned to Egypt by sea and, after concentrating at Cairo, the entire Regiment moved up the Nile to Wadi Halfa. Here it joined the Frontier Field Force, which consisted of one British and one Egyptian brigade and was commanded by General Sir Frederick Stephenson.

The dervishes had gradually built up their strength on the frontier and by December seemed to be on the point of launching their planned invasion of Egypt. Stephenson marched out to meet them and a pitched battle took place at Ginnis on the 30th, the infantry being deployed in line with the Twentieth's squadrons covering their flanks. The enemy attack stalled under concentrated long-range musketry, then broke altogether. At this point the Regiment charged, turning the dervish retreat into a rout; a detachment of one officer and fifty men maintained the pursuit as far as Absarat, fifty miles up-stream. At Ginnis the enemy sustained 800 casualties, the British and Egyptians a mere forty-one. The battle, which removed the immediate threat to Egypt, was remarkable in that it was the last occasion on which the British infantry fought in their traditional scarlet; the Regiment seems to have worn khaki drill throughout the campaign.

The Regiment remained on the frontier for the next two years, then began returning home. By the end of 1887, only one squadron remained and this, commanded by Lt-Colonel Fraser, was transferred at short notice to Suakin, in which Osman Digna was taking a renewed interest. The Suakin garrison was commanded by the then Colonel Herbert Kitchener who, as a Royal Engineer officer, had naturally put the town's defences into good order. Wisely, Osman embarked on a policy of distant blockade rather than direct assault, and established a camp at Handub, some fifteen miles distant. Kitchener personally led a sortie against this on 17th January 1888, the affair coming dangerously close to disaster. The Egyptian infantry who made the attack succeeded in breaking into the camp but were then counter-attacked and had been virtually surrounded when Kitchener arrived on the scene with his cavalry. Fraser's squadron was forced to charge three times before the situation was retrieved and then covered the march back to Suakin, during which Kitchener was himself shot through the jaw. The enemy's casualties at Handub were estimated at 500, the British and Egyptians sustaining fifty. Thereafter, further sorties were forbidden by higher authority, and the squadron returned to Cairo.

The Mahdi had died shortly after Gordon but his expansionist policy was maintained by his successor, the Khalifa Abdullah. In 1889, a dervish army crossed the Nile frontier and, although checked at Argin, continued to advance on Aswan. The squadron was brought forward by river transport and disembarked at Toski on 1st August, riding out at once to shadow the enemy's movements. Two days later it took part in a seven-hour battle in which the dervish force was utterly defeated by the Anglo-Egyptian army, then commanded by General Sir Frederick Grenfell. The battle, which cost the enemy over 1,000 men, a

---

*Osman Digna was a former slave trader who hated the British for ruining his business. His precise antecedents remain obscure, but one interesting rumour had it that he was really a Frenchman named George Vinet!

A Regimental Field Day during the 1890s.

quarter of their strength, ended with a charge and pursuit by the Twentieth and the Egyptian cavalry, in which the dervish commander, Sheikh Wad-al-Nejumi, was killed after his camel had been brought down.

The squadron rejoined the Regiment at Aldershot in 1890, the Twentieth's service in the Sudan being recognised by the award of the battle honour Suakin 1885. In 1895 the Regiment sailed for India, where it remained until 1901.

South Africa

NOBODY'S OWN, in garrison at Mhow in India, must have watched the progress of the Second Boer War with intense frustration as regiment after regiment of regular cavalry, Imperial yeomanry, dominion light horse and mounted infantry were committed to the fray. Ostensibly, Kitchener had more than adequate resources with which to complete his task, yet manning the long blockhouse lines swallowed up troops at a prodigious rate and the demand for reinforcements actually rose throughout 1901. It was, therefore, with some relief that the Regiment received orders to mobilise for active service in November of that year.

By the time the Twentieth disembarked at Durban the war had only five months to run and drives against the elusive Boer commandos had reached a new pitch of intensity. In February 1902, the Regiment took part in the Harrismith Drive, a major operation in Orange River Colony intended to trap the Boers in a small area between the advancing cordon, the Drakensberg Mountains, and the blockhouse lines to the north and south. During this the Twentieth acted as the cordon's beaters, advancing across country with two squadrons up in extended line and ten paces between riders; by coincidence the blockhouse stop-line on this occasion was manned by the 14th Hussars.

The operation was a notable success, for although De Wet's and other commandos managed to slip out of the trap, abandoning most of their cattle and waggons as they did so, that of Lukas Meyer, completely demoralised, surrendered, accounting for 650 of the 778 prisoners taken. In addition, fifty Boers were killed while trying to break out and 25,000 cattle, 2,000 horses and 200 waggons were captured.

Severe blow though the Harrismith Drive might have been, it did not impair the fighting ability of those commandos still in the field. March saw the worst British defeat for two years when De la Rey destroyed a 1,200-strong column commanded by General Lord Methuen. On 2nd April, another column, commanded by Colonel Nixon, was attacked at Strathnairn. Here one troop of the Twentieth formed the rearguard and, although surrounded, fought on until relieved.

Shortly after this peace talks began and the war formally ended on 31st May. The following year the Regiment, which was awarded the battle honour South Africa 1901–02, embarked for a short tour of garrison duty in Egypt, and returned home the following year.

Advance to Contact – 1914

WHEN, IN AUGUST 1914, Sir John French's British Expeditionary Force marched into Belgium, the gap between its right flank and General Lanrezac's French 5th Army was covered by the 5th Cavalry Brigade, consisting of the Royal Scots Greys, the 12th Royal Lancers and the 20th Hussars. Advancing towards the BEF from the north and outnumbering it by three-to-one was General Alexander von Kluck's invading 1st German Army, which formed the outer edge of the giant turning movement which was intended to crush the Allied armies inside a steel fist.

The Regiment, commanded by Lt-Colonel G. T. G. Edwards, sent out patrols and on 22nd August these had several brushes with the German cavalry. Next day, von Kluck blundered into the positions held by the British II Corps at Mons and was stopped dead by the remorseless rapid fire of the infantry. The BEF felt quite capable of staying put, but was forced to withdraw in order to conform with the French. The retreat was screened throughout its duration by the cavalry, who were not greatly impressed by their opponents. This was not simply a matter of long service professionals looking askance at the recalled conscript reservists who filled the German ranks, for the enemy's handling of his own cavalry showed neither dash nor aggression and was, on occasion, totally inept.

At Moy on 28th August, for example, a strong German squadron was wiped out in a brisk counter-stroke by the 5th Cavalry Brigade. The squadron had galloped over a crest some 500 yards distant and had dismounted well down the forward slope, opening fire from a standing position just ahead of its horses, a combination of serious tactical errors for which any British troop leader would have come close to being flayed alive. The brigade's rifles and machine-gun sections at once returned the fire and J Battery RHA sent over a few rounds of shrapnel which stampeded the German horses. The Greys and the 12th Lancers were ordered to make a mounted attack while the 20th Hussars covered their left flank. Many of the Germans surrendered and others were cut down. A horse artillery battery which came to their support was neutralised by the Twentieth and might well have been captured had not wire fences prevented a charge; as it was, the gunners were subjected to accurate rifle fire from a range of only 400 yards and forced to look to their own defence.

Retiring from a less capable enemy was obviously a galling experience for good troops and by 1st September one member of the Regiment, Private Hayhurst, had had enough. He was riding rear point in a patrol commanded by Lt Hall and when this broke contact he observed a group of enemy lancers closing in on its rear. Enraged, he drew his sword and charged single-handed, putting them to headlong flight.

Meanwhile, von Kluck's line of march had begun to diverge from that of his nearest neighbours, von Below's 2nd German Army. By 5th September a 30-mile gap had been detected between the two armies and the BEF was ordered to turn about and march into this while the French also returned to the offensive, the whole sequence of events being later described as the battle of the Marne. The penetration of the gap, held only by a cavalry screen, wrecked the entire German strategy for the campaign. Contact with the enemy was light and the prisoners taken were generally very tired and rather miserable. On 7th

September, Corporal Goring distinguished himself by killing two lancers, retaining their weapons as trophies which he later presented to his family.

By 12th September, the advance had reached the Aisne, beyond which the Germans were frantically entrenching the high ground and stringing wire across their front. The British infantry secured bridgeheads through which it was hoped to pass the cavalry, but the whole valley was being swept with fire and it was soon realised that an attempt to break out would be suicidal. On the 14th, the Regiment, subjected to intense shelling, led its horses in single file across a pontoon bridge at Vailly and then, since its attack was cancelled, was forced to lead them back again; in the circumstances it considered itself extremely lucky that only eleven of its members were killed or wounded.

The fighting on the Aisne set the pattern of trench warfare which was to last for the next three years, but in the meantime both sides attempted to outflank each other at the northern end of the line, where open country existed for a little while longer. This series of movements, known as the Race to the Sea, resulted in the BEF being re-deployed in Flanders for the protection of the Channel Ports and here, between 15th October and 30th November, it fought the First Battle of Ypres. This began as a series of encounter battles in which British and Germans each strove for possession of tactically important features, and ended in the repulse of the latter's frenzied attempts to break through. The Germans, who persisted in employing dense attack formations which formed perfect targets, sometimes call the battle the Massacre of the Innocents, a reference to the complete destruction of several regiments recruited from university students; but it was not merely half-trained idealists who came to grief amid the musketry storm, for when the crack Prussian Guard was pitched into the fray it too was shot to tatters. On the other hand, the Old Army of professionals had been bled white when the battle ended, some of its infantry battalions mustering only fifty men and some of its cavalry regiments less than a full squadron.

After some small scale mounted actions during the early phases, the Cavalry Corps fought on the Messines sector as infantry. Many regard First Yypes as its finest achievement and it received a sincere if involuntary tribute from a captured German officer who remained convinced that the line had been held by machine-gun troops.

The 20th Hussars suffered less severely than some but had their fair share both of action and of casualties. On 15th October, two troops from C Squadron drove the enemy's picquets from the hamlets of Hollebeke and Houthem, while A and B Squadrons dismounted and pushed the German skirmishers back along the line of the Kortekeer Beck. Three days later, B and C Squadrons captured and held the village of Tenbrielen, clearing out snipers as they did so. After this, the horses were sent to the rear and, as the enemy began to intensify his attacks, the Regiment went into defensive positions near Oostaverne; initially, its trenches had to be excavated with cutlery as the entrenching tools had been left behind during the advance to the Aisne. A particularly severe attack on the 30th caused the 12th Lancers to pull back in conformity with the troops on their left and when the enemy came up to occupy their position, the Twentieth's left flank was enfiladed. The Regiment was itself thus forced to retire, the withdrawal being covered by the machine-gun section under Lt McConnel and B Squadron's 4th Troop. The latter, commanded by Sgt Bassinthwaite, who had taken over when his troop leader was killed by a sniper, deliberately sacrificed itself and only one of its men rejoined. The enemy renewed his assault the following day, capturing the village of Wytschaete. At dawn on 1st November, B Squadron, who had comrades to avenge, joined the 12th Lancers in the counter-attack which recovered the village and did considerable execution with the bayonet in the process;

as the Twentieth's historian puts it, the cavalrymen's skill with this weapon might have been 'rough and ready', but the will to win was there and they certainly knew enough to fire a round if it jammed. After this the Regiment had a somewhat quieter time for the remainder of the battle, although it remained in the trenches or in immediate reserve until it ended.

The Years of Deadlock 1915–1917

AFTER FIRST YPRES the trench lines extended from the North Sea to the Swiss frontier and the entire Western Front lapsed into a stalemate in which the powers of defence prevailed. The longed-for breach through which the cavalry could be passed, so restoring mobility to the battlefield, remained not so much an illusion as a mathematical impossibility, for the defence was always able to move its reserves to a threatened sector in far less time than their opponents could traverse the wasteland of shell craters, abandoned trenches and festoons of rusting wire. Throughout this period the Regiment took its turn in the trenches, provided working parties and stood to with the rest of the cavalry as every attempt to break through the German lines failed with horrendous loss of life. In October 1915, Colonel Edwards left to command an infantry brigade, his place being taken by his Second-in-Command, Lt-Colonel G. T. R. Cook.

The British solution to the problems posed by trench warfare, the tank, made its appearance on the battlefield in 1916. The results obtained were initially disappointing, partly because the numbers involved were too few, and partly because the going was often totally unsuitable. The protagonists of the vehicle never lost faith in their idea, however, and on 20th November 1917 the Tank Corps launched a mass assault over hard ground on the Cambrai sector, smashing a yawning gap in the German line. For a few hours a real chance existed for the Cavalry Corps to ride through and out into open country, despite the usual difficulties. This golden opportunity was missed by the Corps commander, Lt-General Sir C. T. McM. Kavanagh, who remained at his headquarters, six miles behind the fighting, from whence he was unable either to visualise or exploit the situation's true potential. By evening the Germans had all but sealed off the penetration and his regiments were, to their disgust, retiring to attend to their mounts.

Kanavagh was not alone in under-estimating the scale of the Tank Corps' success and indeed the whole battle was fought without adequate infantry reserves. On the 25th, the 5th Cavalry Brigade was ordered to produce a composite battalion to which the Regiment contributed five officers and 218 other ranks, and this took part in the fighting for Bourlon Wood, sustaining serious casualties.

The battalion was relieved on the 29th but had hardly rejoined its parent regiments when the enemy counter-attacked the southern sector of the penetration with such force that the forward troops were quickly overrun and surrounded. Few realised it at the time, but they were witnessing the German solution to trench deadlock. This involved infiltration by specially trained storm troops, who bypassed areas of resistance and continued to advance into their opponents' rear areas, closely supported by waves of ground attack aircraft; the technique, which changed the nature of fighting on the Western Front as much as did the tank, relied for its efficiency on constant movement and the reinforcement of success.

The 5th Cavalry Brigade received orders to hold the line near Gouzeaucourt, as this showed signs of disintegrating. Colonel Cook had actually anticipated these and the Regiment was already riding towards the front. Just short of Gouzeaucourt it came under fire and dismounted, taking over some trenches that were thinly manned by Royal Engineers.

A Guards brigade arrived and the Twentieth joined this in a counter-attack, advancing 1,000 yards and inflicting sharp losses on the enemy's storm troops. 'It was a magnificent sight,' wrote a hard-pressed infantry officer who witnessed the action. 'The arrival of the cavalry, fighting as infantry, gave us great encouragement.' Major A. C. Little, the Regiment's Second-in-Command, rallied three infantry companies whose officers had been killed and brought them into the line on the Twentieth's left, so closing a gap which had opened between themselves and the Guards, and for this was awarded a Bar to the DSO he had won at Wytschaete in 1914. During the afternoon an attack by tanks finally stabilised the situation on the Gouzeaucourt sector and the Regiment was relieved the following morning.

The Year of Decision – 1918

THE GREAT OFFENSIVE which Ludendorff launched in the spring of 1918 contained all the elements of the German counter-stroke at Cambrai, to which had been added scientific artillery preparation designed to stun or gas the defenders of the forward zone before the storm troops arrived. By driving directly on Amiens and the Somme estuary, Ludendorff hoped to isolate the British from their French allies and so create the conditions which would ensure a German victory.

When the offensive opened on 21st March, much of the 5th Cavalry Brigade was serving on the St Quentin sector in the dismounted role, and thus lay directly in the path of the German advance. The next eleven days were the most confused in the history of the 20th Hussars, for the enemy's breakthrough was complete and such troops as were immediately available were thrown willy-nilly, irrespective of their origins, into the breach in a desperate attempt to contain the flood. Detachments from the Regiment, both mounted and dismounted, joined improvised battle groups which held vital ground as long as possible or made local counter-attacks. Typical of these engagements was the counter-attack on Rifle Wood, near Amiens, which took place on 1st April. In this a dismounted squadron of the Twentieth, commanded by Captain W. D'Arcy Hall, advanced through enfilade fire to sieze its objective in fierce hand-to-hand fighting and then held it until relieved. The squadron sustained forty-six causalties out of a total strength of 138 and three of its four officers were wounded.

Actions such as this finally halted the German drive, but at a fearful cost. When, a few days after Rifle Wood, the Regiment was brought out of the line, it numbered only thirteen officers and 150 other ranks. Among those killed was Colonel Cook, who had died while commanding an *ad hoc* mounted force known as Cook's Detachment, and command was assumed by Major Little.

The Germans, too, had suffered dreadful losses. Their army had been culled of its best men to recruit the storm troop battalions, most of which had been destroyed. The remaining troops were inevitably of lesser quality and their morale slumped when the victory promised by Ludendorff failed to materialise. As the summer wore on they began to think only of peace and the number of desertions began to rise.

Haig was meanwhile preparing a massive counter-stroke on the Amiens sector. Here, on 8th August, the Tank Corps simply rolled over the enemy trenches, the defenders of which surrendered in droves, and the cavalry passed through. 'We passed numbers of dead Germans, and there were also guns, vehicles and other material that had been abandoned by the enemy,' wrote Captain J. C. Darling, commanding C Squadron. 'Everyone felt that this was indeed a great day and that the attack was the most completely successful one that had yet been made by the British Army. We did not realise then what we know now, that August the 8th was the beginning of the end of the German Empire as a military machine.' There were, however, still Germans willing to make a fight of it, notably the machine-gun teams, who possessed a dedication similar to that of the storm troops. The following day, near Beaufort, C Squadron was forced to make a detour because of wire entanglements and came under intense machine-gun fire before it could reach dead ground, the horses suffering cruelly.

Nonetheless, the advance which began on 8th August continued until the war ended. The Twentieth were frequently detached from 5th Cavalry Brigade for duties with various infantry formations, and on one occasion were attached as divisional cavalry to the US II Corps. By the end of the first week in November, it was clear that the Germans had embarked on a general withdrawal, covered by strong rearguards. The Regiment, under Major G. A. Sanford during the temporary absence of Colonel Little as acting commander of 5th Cavalry Brigade, first rode through the infantry on the 5th, Lieutenant Bland's troop capturing two heavy guns (probably 210 mm howitzers) and forty-nine men in Favril. By the 9th, the Regiment was riding five miles ahead of the infantry. By the 10th, it had reached Clair Fayts on the Belgian border and here its trumpeters sounded the *Cease Fire* at 11.00 am the following morning. At that precise moment a staff officer was having great difficulty in convincing Lieutenant Burt and his patrol that the war had actually ended and that they must not wipe out the enemy machine-gun post they had been stalking. In general, the news was received with quiet relief and a more immediate cause for celebration was provided by the arrival of the ration waggons, as the Regiment had not eaten for two days.

The Last Charge

AFTER SPENDING some time with the Army of Occupation in Germany the Regiment returned home briefly and then embarked for Egypt in 1919 under the command of Lt-Colonel M. C. Richardson, who had led A Squadron for much of the Great War.

The victorious Allies had meanwhile dealt very harshly with Turkey at the Peace Conference, not only dismembering her empire but also dividing Anatolia, the Turkish heartland, into spheres of influence. These latter provisions were too much for patriotic officers like Mustapha Kemal and he led a Nationalist uprising. The Allies sent troops to Constantinople and, in June 1920, the 20th Hussars joined General Ironside's British contingent, which was occupying the Izmid Peninsula.

Shortly after the Regiment arrived, the Nationalists severed communications with the Peninsula by capturing the village of Gebze and blowing up its bridge. A small battlegroup, consisting of the 20th Hussars, the 2/39th Royal Garhwal Rifles, an artillery battery and some Royal Engineers, was sent out to dislodge them on the morning of 13th July.

The Garhwalis had shaken out to assault the enemy's positions in front of Gebze and were watching the artillery's shells bursting among the Turks when the Regiment, 'two squadrons abreast in columns of troops with the third squadron in depth', broke the skyline of a ridge to their left. Only 1,000 yards of good going separated it from the enemy's right flank and Colonel Richardson decided to attack immediately. The Regiment remained perfectly controlled throughout, its trumpeters sounding first the *Trot*, then the *Canter*, and finally the rising cadences of the *Charge*. The artillery ceased firing and, swords fully extended, the Twentieth struck the Turks at full gallop, sending them sprawling or cutting them down. The Regiment rode straight through the enemy, rallied, re-formed, and charged again. The Turks, however, had had enough and were already running as fast as their legs could carry them. The Twentieth's only causalty had been one officer wounded and, well satisfied with the outcome of the affair, they trotted back over the ridge. With the settling of their dust ended a long and honourable era for, although mounted action did take place in Syria during World War II, this was the last regimental charge ever made by British cavalry. If, hearing of the event in some after-world, Shakespeare's centurion had again asked 'Is this well done?' his answer would have been the same: 'It is well done, and fitting . . .'

The Regiment was relieved by the 3rd Hussars in October and, since it already knew its independent life was about to end, transferred 225 of its men to that Regiment. A further 118 were posted to the 11th Hussars and the remainder returned home to amalgamate with the Fourteenth.

The 14th/20th King's Hussars

Squadron leader issues radio orders before an attack.

Mechanisation

THE COMBINED REGIMENT inherited a total of fifteen battle honours earned during the Great War, the 14th Hussars contributing Tigris 1916, Kut-al-Amara 1917, Baghdad, Mesopotamia 1915–18 and Persia 1918, and the 20th Hussars Mons, Retreat from Mons, Marne 1914, Aisne 1914, Messines 1914, Ypres 1914, '15, Cambrai 1917, '18, Somme 1918, Amiens and Sambre. Notwithstanding, an amalgamation is a difficult process and for those involved life is never quite the same again although, despite their very different personalities, the Fourteenth and Twentieth knew each other well, having exchanged drafts of men and cross-posted officers for very many years. In the new Regiment A Squadron remained the 20th Hussars squadron, wearing the Twentieth's badge, while the other sabre squadrons retained the Fourteenth's badges, Headquarters Squadron being manned with personnel from both regiments. The new Regiment was at first known simply as the 14th/20th Hussars, the title King's not being sanctioned again until 1936. The Prussian Eagle made its re-appearance as a badge about 1926, although its use as such was not officially authorised until 1931.

In 1931, the Regiment sailed for Egypt, where it remained for two years and was then posted to India. In 1934, the Inspector General of Cavalry frankly admitted that he could see no future for the horse on the battlefield and requested that his regiments be converted to armour, a process which had, in fact, already begun. The Regiment said farewell to its

General Inspection, Abbassia Barracks, Egypt, 1932.

A mounted parade held at Meerut shortly after mechanisation. Although not easily discernable, the tanks have the Prussian Eagle painted on their turret sides.

troop horses at Lucknow in August 1938, but not to the horse itself, which has always played an important part in its social and sporting life. Shortly after, an instructional team from the Royal Tank Corps arrived with the first tanks, three Vickers Light Mark IIs, and commenced the conversion programme. Tanks continued to arrive slowly but it was not until January 1940, four months after the outbreak of the Second World War, that the Regiment was equipped with a full complement of Light Mark IIs and VIs.

On parade, the Regiment's tanks presented an extremely smart appearance, a prominent black Prussian Eagle having been painted on their turret sides. The four RHQ tanks were named after British castles, the Commanding Officer's being *Windsor Castle.* A Squadron named their vehicles after Regimental battle honours, B after Derby winners, and C chose names with naval associations such as *Iron Duke* and *Sturdee.* The practice of naming vehicles fell into abeyance about 1943.

The Royal Armoured Corps was established in April 1939 to incorporate in one body all those elements within the British Army which were then equipped with tanks or armoured cars. By 1942, the Corps had reached its maximum strength of 104 regiments, including the Cavalry, the Regular and Territorial battalions of the Royal Tank Regiment, a major part of the Yeomanry and a large number of infantry battalions which had converted to armour. Inevitably the experience of these regiments was uneven and diverse. Some fought throughout the war and of these a few had lost all their pre-war officers by 1942 and most of their original members by 1945; others fought one or two actions and were then disbanded to provide replacement crews, or did not get into action until 1945; others again reverted to different roles without ever seeing action in armour. The 26th Hussars, for example, to the formation of which the Regiment contributed a cadre of nine officers and 102 other ranks, were among the last mentioned and were disbanded, a substantial number volunteering for service with the Chindits.* Fortunately, the 14th/20th Hussars were spared the extremes of being either fought within an ace of extinction or being left out of battle altogether. On 16th May 1941, the Regiment received orders to prepare for active service; by coincidence, its first campaign would be fought in Mesopotamia and Persia, over much the same ground covered by the 14th Hussars in the Great War.

*Perversely, the last British armoured regiment to come out of action in the Second World War, and also the furthest from home, was 116 Regiment RAC. They were Gordon Highlanders and fiercely proud of the fact, wearing the Tam o' Shanter in preference to the beret.

The Campaign in Persia 1941

FOR ALL ITS FAULTS, the old Turkish Empire had been a stabilising influence throughout the Middle East and its dismemberment released those pent-up political and religious forces which have rendered the whole area volatile to this day. During the early years of World War II, German agents were also hard at work mobilising anti-British sentiment and in the spring of 1941 Iraq, as Mesopotamia was now called, had witnessed a series of attacks on British bases in the country as part of a *coup* led by the pro-Nazi Prime Minister and four of his generals who called themselves The Golden Square. By the time the Regiment, commanded by Lt-Colonel L. H. S. Groves, disembarked at Basra on 3 July, the dissident Iraqi elements had been defeated, although the overall strategic situation remained potentially serious.

German agents were also at work in the ancient kingdom of Persia, where the security of the oilfields was a matter of the gravest concern to Great Britain. The Shah strove to pursue a neutral path but was clearly subject to a most unwelcome degree of German influence. Nevertheless, it is possible that the *status quo* might have been maintained had not Germany invaded Soviet Russia in June 1941. Such was the scale of the initial German successes that a Soviet collapse seemed likely, the probable results of which would have been to place the Wehrmacht in a position to strike south into Persia, thereby inflicting critical damage on the British war effort. For once the British and Soviet governments found common cause and decided again to implement the *entente* of 1907 which, it will be recalled, divided Persia into respective spheres of influence. The Shah having refused to accede to a request that all German nationals be asked to leave his country, an invasion by the Allies became inevitable.

On the British sector this would involve the 8th Indian Division, whose objectives were Abadan, Khorramshahr and Ahwaz, and Major-General William Slim's 10th Indian Division, which would force the Pai Tak pass and advance through Kermanshah to Hamadan, thereafter effecting a junction with those Russian troops advancing south from the Caspian. Also under Slim's command were the 2nd Indian Armoured Brigade, at this stage consisting solely of 14th/20th Hussars, and the 9th Armoured Brigade, whose three regiments (the Household Cavalry, the Warwickshire and the Wiltshire Yeomanry) were armoured in name only and equipped with nothing more substantial than lorries.

The Regiment now possessed a full complement of Vickers Light Tanks Mark VIB and these were shipped by rail from Basra to Baghdad, while the remainder of the vehicles moved by road. Whatever had been said by an earlier generation regarding the unpleasantnesses of soldiering in 'Messpot', the reality was every bit as bad. During dust storms the vehicles' paintwork was scoured by flying sand until the angles gleamed with a silvery sheen. Baghdad during the hot season was as unbearable as ever, it being quite impossible to enter the tanks while the sun was at its height. Heat-stroke claimed several victims and actually proved fatal in the case of the adjutant, Captain R. J. W. McAllen.

Slim had considered his attack on the Pai Tak pass very carefully. He anticipated that the pass itself would be held in strength and therefore divided his force into two columns the first of which, led by the Regiment less C Squadron, would advance up the pass while the second, led by C Squadron, would cross the frontier a little way to the south and penetrate the Persian hinterland through Gilan and Shahabad, thus turning the enemy's position at Pai Tak. As to the degree of resistance which might be expected, the Persians remained something of an unknown quantity. On the one hand the Persian Army was not in the same class as the British, yet on the other it held excellent positions and had just been re-equipped by Germany with first-class weapons from the captured Czech arsenal; these included a 47 mm anti-tank gun which was capable of sending an AP round straight through a Vickers Light without the slightest difficulty.

The advance of the main column began at midnight on 24th/25th August and by dawn the frontier town of Qasr-i-Shirin had been secured without difficulty. The column then advanced unopposed throughout the day to the foot of the Pai Tak pass, where it halted. With B Squadron in support, the Reconnaissance Troop pushed on up the road towards the summit but soon came under heavy and sustained fire and were pulled back.

Slim decided against an immediate assault and the following day he ordered most of the column to join the force which was advancing on Gilan, leaving an infantry brigade to watch the pass. This lateral movement involved a difficult march through a gorge but after five hours of struggling with the rocky terrain the vehicles broke through. The welcome extended by the Warwickshire Yeomanry was unexpectedly warm – believing the column to be Persian they opened fire, fortunately without serious consequences, although they did manage to shoot one troop leader's microphone out of his hand!

The Gilan column, commanded by Colonel Pocock, had made good progress and had captured Gilan village on the morning of the 25th. Some way beyond, however, the two sections of the Reconnaissance Troop which, under Lieutenant Talbot, had accompanied the column, came under heavy fire from concealed positions on a ridge to the south-east and were pinned down. Talbot reported the situation and Pocock ordered Captain P. F. W. Browne, commanding C Squadron, to take two troops up the road and draw the enemy's fire, so enabling the artillery to register the position of the Persian guns.

The troops of Lieutenants Donovan and Campbell were detailed for the task and with them Browne set off to trail his coat in front of the excellently concealed Persian positions, soon succeeding in drawing the desired response. In ideal conditions he would have employed the Vickers Light's only asset, which was its speed; unfortunately, the going was extremely rough on the road and impossible off it, so that the squadron was forced to proceed in line ahead and at walking pace, surrounded by bursting shells. On the return leg of the journey Browne spotted an enemy gun which he knocked out, and then two more, whose crews he chased off with his machine gun. Incredibly, only one tank sustained serious damage and although the turret crew were killed the driver managed to bring it out of action.

The British artillery now went to work and Pocock decided to send the squadron up the road again with a view to pursuing the enemy when they broke. This was a slightly optimistic appreciation, for at this stage the Persians were still holding firm, and the tanks were subjected to a hail of fire as they moved slowly forward. Donovan, swinging off the road to get a better shot at an anti-tank gun, became ditched in a wadi and, seeing his predicament, the enemy concentrated their fire on his vehicle. Browne told him to stay put and arrived on the scene in an open tracked carrier, driven by SSM Senior, accompanied by Campbell, who manoeuvred his tank so that it covered the recovery operation. While shells burst nearby and bullets kicked up the dust all around, Browne dismounted and attached a

Lt-Colonel P. F. W. Browne, DSO, MC.

tow line to the carrier, which pulled Donovan clear at the second attempt. At this point the carrier ran out of fuel. The tow line had to be attached again and this time Donovan towed Browne out of a tight spot, covered by Campbell. It was now clear that the enemy could not be ejected by nightfall and the squadron was withdrawn into leaguer.

At first light, Donovan moved up the road and once more came under fire, his tank shedding a track close to the spot where it had become ditched the previous day, and again Browne and Senior arrived to tow him clear. The Persians, however, were pulling out, and a slightly later patrol by Campbell confirmed that they had gone. The column continued its advance and at Karind met the 21st Indian Brigade driving down the Pai Tak pass, the defenders of which, seriously alarmed that Pocock's force would cut them off, had abandoned their positions after being attacked from the air.

Imperial War Museum

A Light Tank Mark VI of A Squadron drives past newly surrendered prisoners of war, Persia, 1941.

Imperial War Museum

The Regiment entering Kermanshah, Persia, 1941.

Slim's combined force pressed on along the main road but was halted by a strong Persian position on a ridge at Zibiri. This blocked all further progress and a set-piece attack would be required to take it. The Regiment had actually deployed to support this when at 1100 on 28th August a Persian officer arrived with a flag of truce to say that the Shah had ordered his troops to cease fire.

It remained only to complete the journey to Hamadan in the shortest possible time, since it was feared that the Russians were quite capable of enlarging their own sphere of influence beyond the agreed boundaries, given the chance. This involved climbing several more passes up to 9,000 feet in height and required the most skilful driving particularly as the Vickers Light, if allowed to run away, suffered from a fault known as reverse steering which could send one in precisely the direction opposite to that in which one wished to travel. With sheer drops of hundreds of feet bordering the narrow road for much of the way it is safe to assume that drivers kept their mounts under the strictest possible control.

Contact with the Russians was established on the Hamadan - Kasvin road, their attitude predictably varying between the jovial and the deeply suspicious. Fortunately, having troubles enough of their own, they were prepared to abide by the letter of the agreement with their allies. Shortly after, the Shah abdicated in favour of his son and the process began of rounding up the large numbers of German and Italian agents throughout the country.

The Regiment's first campaign in armour is of some historical interest since it provides the only instance of the Vickers Light, unsupported by heavier tanks, influencing the course of a major engagement. The tank had been designed with the North West Frontier in mind, where small arms fire was usually the worst that could be expected. Experience in France and the Western Desert had revealed that it was frighteningly vulnerable to heavy weapons and really had no place at all on the modern battlefield. It had certainly broken the enemy's resistance at Gilan, but the real reasons for this lay in the courage and determination of Browne and his crews, in recognition of which Browne was awarded an immediate Military Cross.

The campaign had the effect of securing the Persian oil for the remainder of the war and also provided a direct route from the Persian Gulf to the Caspian along which the Western Allies passed five million tons of desperately needed military equipment to Russia; this is sometimes forgotten, notably by the Russians themselves.

The Middle East 1942–1944

IN OCTOBER 1941 the Regiment returned to Iraq and was stationed near Mosul. Here, during the spring of 1942, its light tanks were replaced by Stuarts, which were in approximately the same class as the British cruiser series, being armed with a 37 mm gun and several machine guns, as well as being nimble and reliable.

Meanwhile, the Reconnaissance Troop, commanded by Captain Brian Tayleur, was engaged in filling in the numerous gaps which existed in the only maps of the area. Their travels took them over the north Persian frontier and into Russian controlled territory, where their reception was rather mixed. Tayleur called on a cavalry squadron whose officers introduced him to the time-honoured Russian military custom of drinking oneself insensible, and then rashly provided a display of riding and swordsmanship for their British guests. The squadron commander's first attempt to mount resulted in his taking a header down the offside of his horse. His second was more successful but, as Tayleur noted with interest, 'Unfortunately, he could not control sabre, horse and vodka and he disappeared into the trees in the distance, clinging to his horse's neck!' The squadron's political commissar, a large grumpy man, had no such illusions regarding hospitality. His mood swung uncertainly between brooding on the apparently deadly insult posed by the British presence to his own dignity and that of Mother Russia, begging lifts, and threatening to have everyone locked up. Drink inclined him progressively towards taking the last mentioned course of action, but the Troop beat him to his own roadblock and waved him goodbye in the rudest way possible.

In June the Regiment went back to Kermanshah in Persia, so avoiding the worst of the hot weather, and in November moved to Shaiba in southern Iraq. In January 1943 the Stuarts were exchanged for Grant medium tanks, which mounted a 75 mm main armament in a sponson on the right side of the hull and a 37 mm gun in a top turret. At this period the Regiment formed part of Major-General R. E. Wordsworth's 31st Indian Armoured Division, which consisted of:

252nd Indian Armoured Brigade (Brigadier G. Carr-White)
14th/20th King's Hussars
4th Duke of Cambridge's Own Lancers (Hodson's Horse)
14th Prince of Wales' Own Cavalry (Scinde Horse)
1st/4th Bombay Grenadiers
43rd Gurkha Lorried Infantry Brigade (Brigadier A. M. Barker, OBE)
2nd/6th Gurkha Rifles
2nd/8th Gurkha Rifles
2nd/10th Gurkha Rifles
Divisional Troops
13th Duke of Connaught's Own Lancers
15th Field Regiment, RA
144th Field Regiment, RA
79th Anti-Tank Regiment, RA
Madras Sappers and Miners

Officer and troopers, 14th/20th King's Hussars, Iraq, 1942.

The fact that many regular regiments had already spent a considerable period abroad when war broke out was particularly hard on men whose families remained in the United Kingdom, and a new scheme called Python was introduced which entitled them to repatriation. This meant that the Regiment now lost many of its most experienced soldiers. Luckily, their places were taken by former 7th Queen's Own Hussars who had been battle-hardened in the Western Desert and in Burma, and by Lancashire Fusiliers, three battalions of this regiment having been converted to armour. In February 1943, Lt-Colonel Groves returned to the United Kingdom and command was assumed by Lt-Colonel H. A. R. Tilney.

The 31st Indian Armoured Division moved to Quayara, south of Mosul, in May and from thence to Burg-el-Arab, west of Alexandria, in October. The Regiment exchanged its

Grants, which were difficult tanks to get hull-down because of their low-slung main armament, for Shermans, in which the 75 mm gun was mounted in a conventional turret.

In April 1944, the Division was sent north and spent most of that year training with infantry formations for mountain warfare in Syria, Lebanon and Palestine. In May, the Regiment's personnel were ordered back to Alexandria at short notice, taking only seventy-two hours to complete the journey from Mount Hermon, draw a complete vehicle establishment and render it serviceable. The reason for this sudden re-deployment was the mutiny of a Greek brigade at Burg-el-Arab; this collapsed as soon as the Regiment closed in on its barracks. After this brief period of excitement the Regiment returned to Syria, taking part in exercises along the Turkish frontier the following month. In July came the shattering news that the 31st Indian Armoured Division would never see active service as such.

It was not in the Regiment's nature to stand apart while great events took place, yet for three years it had been forced to do so. Its efficiency was not in question, for in 1944 the Commander-in-Chief Middle East commended the Regiment for this very quality as well as praising it for its boundless patience, while the Vehicle Inspectorate Board awarded it the classification of Excellent, a grading never previously attained by a unit in the field. Naturally, the frustrations of the period led to discussion as to why the Regiment had not been required to serve in a more active theatre of war. Some of the older hands recalled that during the pre-war tour of duty in Egypt one officer had had a blazing row with a certain Lt-Colonel Bernard Montgomery at a sporting event. It was said that Montgomery had never forgotten the incident and that he disliked the Regiment and was averse to having it under his command. This is probably a little unfair, for in his time the future Field Marshal was involved in pointed disagreements with a great many people, and had he applied such a yardstick the ranks of the Eighth Army might well have been rather thin. The truth is that a strong armoured presence in the oil producing countries of the Middle East was a vital necessity, not simply to protect the oilfields for Allied use, but also to prevent a German penetration of the area through the back door. The Wehrmacht's 1942 offensive in Russia came frighteningly close to achieving this latter aim with its advance deep into Caucasia, and would undoubtedly have spilled over into Persia and Iraq had not Hitler developed his insane fixation on Stalingrad. The immediate danger did not recede until 1943, but Turkey still provided an alternative route for an Axis penetration of the Middle East and continued Turkish neutrality could not be taken for granted until the following year. By then the Regiment was far removed from the nearest seat of war.

If the fortunes of war had dealt unkindly with the Regiment by placing it in the wrong place at the wrong time, they were about to change. The internal security situation in Palestine had already begun to deteriorate and in December the Regiment converted briefly to armoured cars. Simultaneously, it was advised that it would be relieved shortly by the 3rd King's Own Hussars and would then embark for Italy as an armoured regiment.

Italy 1945

THE REGIMENT sailed from Haifa on 17th January 1945 and landed at Taranto four days later. From here it moved north onto the Florence sector of the line, coming under the command of XIII Corps as an independent armoured regiment covering the junction of the Eighth Army on the Allied right with the US Fifth Army on the left. For the moment a hard winter and lack of resources had brought operations to a standstill but during the next few weeks squadrons were able to shoot themselves in against live targets.

Meanwhile, preparations were being made for the Allied spring offensive which was to end the war in Italy. The previous autumn the Eighth Army had fought its way through the Gothic Line and reached the Lombardy Plain. This was not the attractive tank country it sounded, for much of the ground was marshy and the remainder was cut up by numerous water obstacles; again, there were few good roads and it was a simple matter for the enemy to mine them or demolish bridges. Nonetheless, General Sir Richard McCreery, commanding the Eighth Army, was determined to fight a mobile battle and to this end had been supplied with the same sort of special purpose armoured vehicles that equipped the 79th Armoured Division in North-West Europe. Unfortunately, no such organisation existed in Italy and the Eighth Army was forced to convert existing units to this role. Thus in March, A Squadron exchanged its tanks for Kangaroo APCs, which were normally turretless Shermans but in this case were Unfrocked Priests (i.e. the Howitzer Motor

Imperial War Museum

The Regiment moves up to the front, Italy, 1945.

Carriage M7 Priest from which the gun had been stripped), crewed by a driver and commander/radio operator with the infantry riding in what had been the fighting compartment. A Kangaroo squadron at full strength could lift an entire infantry battalion into action.

With pleasure the Regiment learned that during the coming offensive it would form part of a battlegroup based on the 43rd Gurkha Lorried Infantry Brigade, with which it had trained extensively in the Middle East and which, under Brigadier Barker, had been serving in Italy since the previous autumn. Other units included in the battlegroup, which would operate under the command of II Polish Corps, were the 2nd Royal Tank Regiment, 25th Field Regiment RA, and R Battery 15th Field Regiment RA; in immediate support were the 5.5-inch howitzers of the Duke of Lancaster's Own Yeomanry, a Polish tank destroyer battery and an assault engineer troop.

The Allied plan called for the Eighth Army, led by V British and II Polish Corps, to cross the River Senio and then strike north-west towards, respectively, Ferrara and Bologna; simultaneously the US Fifth Army would advance due north to Bologna, the combined effect being to trap most of the Tenth German Army inside a pocket.

The offensive began on 9th April but Barker's Brigade Group was not involved for several days, its task being to convert the break-in battle into a breakthrough. The Senio was crossed by the Poles, and then the Santerno, with Gurkha assistance. On the night of 15th/16th April, Lt-Colonel Tilney was ordered to capture the town of Medicina, which lay some ten miles ahead and was an important road junction, using a regimental battle-group consisting of the 14th/20th Hussars (with A Squadron 2 RTR under command) and the 2nd/6th Gurkha Rifles riding in the Regiment's Kangaroos. This operation resulted in one of the fiercest actions in the Regiment's entire history.

Medicina and Beyond

FEW UNDER TILNEY'S command had any illusions that the task ahead would be easy. The Germans knew that their country was staring defeat in the face, yet their philosophy was that by denying the Allied armies in Italy access to the Balkans they were enabling their comrades on the Eastern Front to hold off the hated Russians just a little longer while some sort of understanding was reached between their own government and the Western Powers. This might have been the most wishful thinking, but it was the best most Germans could hope for and they fought hard to achieve it. In direct opposition to the Regiment was the 1st Parachute Division, tough, experienced, full of fight and probably the best formation in the German Tenth Army. In support were elements of the 26th Panzer Division and part, at least, of Heavy Tank Battalion 504, which still retained a few of its Tigers.

The battlegroup moved off at first light with A Squadron 2 RTR, commanded by Major P. S. Hunter, in the lead. 'The enemy was definitely in retreat, but rearguards were putting up strong resistance in well-chosen positions. The ground in this area is cut up by a series of rivers, canals and ditches and heavily planted with tall vines. At intervals there are unexpected patches of completely open ground, across which an enemy anti-tank gun could kill as the tank emerged suddenly from the vines.

'At 6 am A Squadron led across the Sillaro, which although practically dry had very steep banks. Two Arks had been put in one on top of the other, but the ramps were still at an extremely sharp angle.* The leading tank slipped and the driver stalled his engine leaving the tank at a sideways angle half-way up the exit ramp. At this moment an accurate enemy salvo arrived slap on the crossing, but the driver managed to straighten up and drive forward off the ramp.'

A second tank crossed but in so doing carved up the exit bank to such an extent that others were unable to follow unaided. For a while it was necessary to resort to the dangerous and time-consuming process of towing tanks off the Arks and up the bank. At length an armoured bulldozer arrived and after the exit had been levelled the Regiment crossed.

Contact had been lost with the Polish troops who should have been on the left and Tilney decided to use A Squadron 2 RTR as a flank guard while the main body of the column pressed on along the road to Medicina. Hunter's squadron advanced with two troops up. 'At La Palazza, the leading tank bogged in a deep ditch. A German armed with a bazooka who crawled up the ditch to snipe the tank was killed before he could fire. An SP gun flashed out of the houses, fired twice at one tank, missed it and disappeared northwards. No 1 Troop Commander (G. C. Sansom, MC) by following in its tracks was able to cross the ditch and throughout the whole day the track marks of this troop of hostile SPs, which paid us much attention, also saved us much reconnaissance time in finding crossing places.'

*The Ark was a Churchill tank chassis with steel decking and folding ramps fitted to the top of the hull. The vehicle was driven bodily into an obstacle such as this and, if the situation required it, as it did here, a second Ark would be driven on top of it to provide the necessary height.

Meanwhile, the Regiment was advancing along the main road, led by C Squadron under the command of Major P. F. W. Browne, MC, who had so distinguished himself at Gilan. The road was un-mined and although it was under artillery fire a good pace was maintained. However, the last major water obstacle before Medicina, the Scolo Sillaro, was heavily defended and here C Squadron became involved in a fierce battle with tank destroyers and infantry armed with Panzerfaust anti-tank weapons. This resulted in two Shermans and a Stuart from the Reconnaissance Troop being knocked out and two further Shermans damaged. The Germans lost one tank destroyer, captured intact, some fifty dead and wounded and about the same number taken prisoner.

During the action a Gurkha company had dismounted and doubled forward, securing a bridgehead around a possible crossing site. Medicina was now in plain view some two-and-a-half miles distant and C Squadron was sending HE shells into the town. Unfortunately, no Arks were available and no other means of crossing the Scolo Sillaro existed. At this point there occurred one of those small incidents upon which the success or failure of major operations depend. The 2nd New Zealand Division had been advancing parallel with the Regimental Group and on its right, and from this direction an armoured bulldozer appeared, scooped out entry and exit points in the banks, then disappeared whence it had come.

The Regiment was soon across and closing in on Medicina. Since tanks are extremely vulnerable when fighting without close infantry support in built-up areas, Tilney was about to order the Gurkhas to dismount and go into the lead when Brigadier Barker arrived, leaving him in no doubt that he wished the tanks to lead the attack. This was contrary to all the established rules of infantry/tank co-operation and the implications required serious consideration.

At such moments a regiment's inherited experience plays a vital, if unconscious, role. Hervey would have understood; so too would Taylor, Brotherton, Havelock, Gall, Prettejohn, Hewitt and all the others. So too did Tilney. Had a mistake on the scale of Ramnuggar been made? It did not matter, for the Regiment had been given a job to do and, whatever the consequences, it would do it. 'Get your whips out and go like hell!' he tersely ordered Browne. Then, knowing that C Squadron was already below strength as a result of the fighting at the Scolo Sillaro, he sent in the 2nd/6th Gurkha Rifles in A Squadron's Kangaroos, escorted by B Squadron.

Medicina was held by part of the 26th Panzer Division, which had been in Italy since July 1943 and was a thoroughly experienced formation.* Also present was at least one platoon of Heavy Tank Battalion 504. Thus far, the nature of the fighting in Italy had followed a definite pattern, particularly in the mountains. Now, the sheer weight and speed of the Regiment's attack presented so radical a departure from that pattern that the Germans were, for the moment, caught on the wrong foot.

Browne followed another Regimental tradition by choosing to lead the attack himself. Guns blazing, he broke into the town to be confronted by a tank destroyer. This was engaged and set ablaze, leaving just sufficient room for his tank to squeeze past. He immediately found himself staring down the black muzzles of two Eighty-Eights,

---

*The division's order of battle at this period was: Armoured Recce Bn 26; Panzer Regiment 26 (two bns); Panzer Grenadier Regiments 9 and 67 plus a Reinforcement Grenadier Regiment (two bns each); Armoured Artillery Regiment 93 (three battalions); Army AA Bn 304; Tank Destroyer Bn 93; Signals Bn 93; Armoured Engineer Bn 93. Because of the tank shortage Panzer Regiment 26 was equipped with assault guns as well as Panthers. The division surrendered near Bologna shortly after Medicina.

The battle for Medicina. Major Browne's tank is edging its way past the knocked-out tank destroyer and has opened fire on the Eighty-Eight on the left of the picture.

positioned on either side of the street; in normal course of events these should have blown the Sherman apart, but the co-axial and hull machine guns cut down their crews before they could open fire. Shot at from every door and window, Browne ground on over the rubble as far as the corner of the station square, where his vehicle was hit and set on fire by a Panzerfaust. Trooper Burt, the gunner, was killed, while Browne and the operator, Sergeant Evans, were badly wounded. Scrambling clear, Trooper Armstrong and L/Cpl McGregor, respectively the driver and hull gunner, stormed the house from which the Panzerfaust had been fired, killed most of its occupants and then returned to lift Browne and Evans to safety. Meanwhile, in the tank behind SSM Long was shot dead in his cupola while giving covering fire with his machine carbine.

Running back down the street, McGregor and Armstrong found Captain D. A. Heath's tank just beyond the burning tank destroyer, and advised him of the position. The tank destroyer's ammunition had begun to explode and the rubble from a collapsed wall now blocked the way past. Assuming command, Heath, who was C Squadron's Battle-Captain, sent a troop by an alternative route into the square, followed by a Kangaroo troop.

Here Corporal Plumley's tank was hit by a Panzerfaust which killed the gunner and wounded every other member of the crew save Trooper Harper, the hull gunner; the German responsible was promptly despatched by a Gurkha kukri. Everyone, except Plumley, had managed to leave the fiercely blazing vehicle but Harper, seeing the Corporal's predicament, climbed back aboard and managed to get him out just before the ammunition exploded.

The enemy's Panzer Grenadiers were good troops but they were no match for the Gurkhas, who chased them from cellar to attic and then across the roof-tops, their fearful yells of 'Ayo Gurkhali!' mingling with the dying screams of their victims.

By now the streets of Medicina were choked by C Squadron's Shermans, A Squadron's Kangaroos, wrecked enemy equipment and the rubble of collapsed buildings. In consequence B Squadron's tanks were unable to get right into the town but gave what close support they could to the Gurkha companies. In the meantime, Tilney had decided to form an operational reserve based on A Squadron 2 RTR, less the latter's No. 1 Troop, which entered fully into the spirit of the engagement and 'was too fully occupied shooting up enemy-held houses at point-blank range, and co-operating with a platoon of Gurkhas against barricaded and sandbagged houses. Wireless messages did, however, eventually penetrate through the fog of battle, and No. 1 Troop unwillingly withdrew.'

Few battles have been fought at such a sustained level of fury. It had been 1945 hours when C Squadron broke into the town and with the light fading steadily the fight resembled a scene from hell itself as demonic figures hunted each other against a background of blazing buildings to the accompaniment of continuous automatic fire and the thump and flash of tank guns. By the time darkness finally fell the only Germans remaining in Medicina were prisoners. The enemy's losses amounted to 159 killed and wounded, 80 captured, two 88 mm guns, five self-propelled guns of various types, three Panthers and one Tiger. From start to finish the Medicina operation had cost the Regiment five tanks destroyed and two seriously damaged, five killed and 23 wounded.

Four immediate awards were made for gallantry. Browne received the DSO and Sergeant Hall, who had continued to fight from his severely damaged tank at the Scolo Sillaro, the DCM; McGregor and Armstrong were awarded the MM. Captain Heath later received the MC, as did Lieutenant K. T. Brailey, who commanded the first Kangaroo troop to penetrate the square. Medicina remains one of the Regiment's most treasured battle honours and its anniversary is celebrated yearly; it also marked the start of a long-standing association with the 6th Gurkhas Rifles.

The following morning the Brigade Group, now operating directly under the command 2nd New Zealand Division, resumed its advance on Bologna. The enemy was dug in along the Gaiana Canal, a water obstacle more formidable than any thus far encountered, and put up the most determined resistance throughout the day in the face of repeated and costly Gurkha attacks, for which the tanks provided direct gunfire support. During this difficult battle, A Squadron's Kangaroos distinguished themselves by bringing forward ammunition and lifting out the Gurkha wounded under heavy fire. The position did not fall until after dusk on the 18th and then only after the most thorough artillery preparation and the arrival of Churchill Crocodile flamethrowers, the effects of which horrified even their own crews. The Germans' stand here enabled many of their units to slip out of the trap which was closing round them but the fact remained that the Tenth Army had been roundly beaten and was now in full retreat. Bologna was entered unopposed on the 21st and the Allied advance continued to the Po and beyond.

After the battle at the Gaiana Canal the Brigade Group was withdrawn briefly to Medicina for rest and re-fitting and here Colonel Tilney, who was unwell, handed over command of the Regiment to Major R. J. Stephen. C Squadron, now commanded by Major P. F. S. Clifford, received orders to convert to Kangaroos and completed the task in thirty hours.

The Germans never succeeded in rallying and surrendered on 2nd May. Indeed, after its period of re-organisation the Regiment was hard pressed to come up with them again, having repeatedly to wait its turn at river crossings. However, having crossed the Adige during the night of 1st/2nd May it finished the war in fine style with a dashing non-stop advance of 130 miles which ended the following noon at Padua. During the few months that it had been with the Eighth Army it had earned the battle honours Bologna, Medicina and Italy 1945.

Other Aspects of the Regiment at War

TO A GREATER or lesser degree the Regiment was represented on all the major battlefronts involving the British and Commonwealth Armies. Some mention has already been made of the 26th Hussars, many of whom served as Chindits in Burma and, again, a number of those who were repatriated from the Middle East under the Python scheme continued their active service in the Royal Armoured Corps after their period of leave.

Inevitably, many members of the Regiment were detached for Staff and other duties and the careers of some of these are of especial interest. Captain A. J. C. Stanton joined the Regiment in 1937 and later became adjutant of the 26th Hussars. In 1943 he was serving with the Australians in New Guinea when he was killed while bringing in a wounded man under fire. Another officer who did not survive the war was Major G. B. Walker who had left the Regiment as a subaltern in 1937 but had joined the 4th County of London Yeomanry in June 1939. Walker became his regiment's adjutant and was commanding a squadron at Sidi Rezegh when he received wounds from which he subsequently died. He had been recommended for the MC but it was not the custom in those days for the award to be made posthumously.

During the fighting against the Vichy French in Syria, Major G. L. Sullivan (1932) was serving with the Trans-Jordan Frontier Force and led a mounted attack, the last 14th/20th Hussar ever to do so. He later served with the East Riding Yeomanry in North-West Europe but was wounded and captured in Holland. The Germans did not find him a model prisoner for he escaped three times, on the last occasion successfully, and was awarded the MC.

Lieutenant-Colonel D. A. H. Silvertop (1933) served as Brigade Major of 4th Armoured Brigade during the fighting in the Western Desert, where he won the MC. He commanded the 3rd Royal Tank Regiment in Normandy, demonstrating his thorough grasp of armoured warfare both there and during the advance through the Low Countries. On 4th September 1944, with the aid of an extremely brave Belgian engineer named Robert Vekemans, he succeeded in capturing the dock area of Antwerp intact, an event which undoubtedly shortened the war in Europe and earned him the DSO. Tragically, he was killed later the same month in Holland. Nonetheless, his achievement was commemorated by the citizens of Antwerp, who named a street after him, and in 1968 a plaque recording the event was unveiled by a representative of King Baudouin, this bearing the badges of his own and his adopted regiment.

The Regiment Since 1945

SINCE THE END of the Second World War the Regiment has performed every role that the Royal Armoured Corps has to offer and has also been closely involved in developing the technology required in a world where weapon systems and battlefield techniques change at a rate which has no parallel in military history.

In two respects, however, the 14th/20th King's Hussars remain completely unchanged. The first is that in spirit and philosophy they are a cavalry regiment, no matter how mounted. The second is that they are a close family. This term is not applied simply to serving members, but also to those who have concluded their active service, and in turn to their families. Again, while sons might follow their fathers and grandfathers into the service, Hawks also marry the daughters, sisters and nieces of other Hawks. Some, like the late Sergeant V. C. Coles, spend virtually the whole of their lives within the Regiment; born in married quarters in 1908, 'Nifty', as he was known, enlisted in 1923 and did not retire until 1962, by which time he had been awarded two bars to his Long Service and Good Conduct Medal.

In July 1945, the Regiment converted to Staghound and Greyhound armoured cars and moved the following month to Chiavari, near Genoa, where it remained until February 1946, when it handed in its vehicles at Milan and entrained for Germany. Here it was stationed first at Luneburg as Reconnaissance Regiment of the 43rd (Wessex) Infantry Division, and then moved to Wuppertal in June to become Divisional Regiment RAC of the 53rd (Welsh) Infantry Division.

The function of the Divisional Regiments RAC was to provide mobile anti-tank defence for their parent formations, and to this end they were equipped with tank destroyers, which were simply a tank chassis on which an anti-tank gun had been mounted inside an open-topped superstructure of armour plate. Several types existed, the Regiment being equipped with the Archer, which consisted of a 17-pdr gun on a Valentine chassis, and was probably the least satisfactory member of the entire family; the long gun, which had a very limited traverse, projected over the engine deck and this meant that the vehicle had to be reversed into its firing position, while the driver was forced to abandon his seat in action, as the breech recoiled straight over it! The tank destroyer concept, while undeniably useful in its day, was one which had provoked controversy within the Army. The RAC had argued that anti-tank guns were the responsibility of the Royal Artillery, and the latter had retorted that tracked AFVs should be manned by the RAC. The Gunners had, on sufferance, performed the role during the war, but had quickly divested themselves of it afterwards.

In January 1947, the Regiment adopted the crossed-kukri arm badge at the request of the Brigade of Gurkhas, thus commemorating its long wartime association with the 43rd Gurkha Lorried Infantry Brigade, and in particular the Medicina operations, this being subsequently approved by the King. In November, the 14th/20th returned to the United Kingdom for the first time since 1931 and for the next three years assumed the task of an Army Basic Training Unit (ABTU) at Catterick. This was an extremely busy period, with fresh intakes of Regular and National Service recruits arriving every fortnight, all of whom had to be turned from civilians into disciplined, trained crewmen before being despatched to their own regiments. In total, the Regiment was responsible for the initial training of 1,732 Regular soldiers and 6,167 National Servicemen, while 276 potential officers were passed on to Sandhurst and Mons OCS for completion of their training.

The Band, which in 1941 had become a mortar troop, was re-formed at Catterick in October 1947, unfortunately without the kettle-drums of either the Fourteenth or the Twentieth, no trace of which have ever been found. This loss was a particularly sad discovery, for the beautifully worked drum banners provided the only way in which a light cavalry regiment could display its battle honours until guidons were reintroduced. It was doubtless with this in mind that an unusual Trooping ceremony was held in 1950, when three Peninsular War guidons belonging to the Fourteenth and Twentieth Light Dragoons were paraded along the Regiment's ranks.

On concluding its tour as an ABTU, the Regiment moved to Crookham in January 1951 and became Divisional Regiment RAC of the 3rd Infantry Division. The RHQ and SHQ Troops were equipped with 75 mm Cromwells, but the Regiment's principal weapon was again the Archer; this was, in fact, the last occasion on which it was required to perform this role, since gun-armed tank destroyers could no longer match the long-range firepower of the tanks they were intended to kill, and were being slowly withdrawn from British service.

The culmination of the 1951 training season was Exercise 'Surprise Packet', the last major exercise to be held in the United Kingdom. This involved a running battle between the 3rd Infantry and 6th Armoured Divisions across an area which stretched from Norfolk to Salisbury Plain, causing a degree of unavoidable damage to civilian property on route; one of the more interesting sights to confront the then Lieutenant J. M. Palmer was one of the opposition's Centurions, deeply embedded in the wall of a pub!

In February the following year, the Regiment sent a contingent to line the route during the funeral of the much-loved King George VI, and that October it embarked for Libya, where it was to remain for the next three years, being stationed at Sabratha near Tripoli. Here it became the armoured regiment of the 1st Infantry Division, which was actually based in Egypt. However, since the Canal Zone offered minimal training facilities, it was decided that the Regiment would be better employed in Libya, whence it could be lifted quickly forward in the event of Anglo-Egyptian relations deteriorating even further than they had already. During this period it was equipped with Comets, replaced after the first year by Centurions.

Libya was a good posting which offered some interesting soldiering as well as equitation and the chance to visit Malta on local leave. At Sabratha, it was possible to visit the magnificent Roman remains, Leptis Magna and other archaeological sites, and the desert itself represented a challenge to the adventurous which the Army as a whole has never been slow to take up.

As well as carrying out the usual round of training the Regiment was closely involved with the Fleet Air Arm, providing camouflaged targets which the aircrew would attempt to identify and 'strafe', or carrying out desert air rescue exercises. One of the latter, commanded by Major G. L. Sullivan, penetrated as far south as the Fezzan in December 1952, spending Christmas with the French Foreign Legion at Fort Leclerc, which dominated the little town of Sebha. Technically, Sebha lay well inside Libyan territory and the French, who belonged to the Chad garrison, had no business there, although no one was greatly concerned about this. The Legion's hospitality was as formidable as its reputation

A Centurion wades ashore during an amphibious exercise in Libya, 1953.

and, whenever possible, contact was renewed between the two units. The Legionnaires themselves were a hard-drinking lot and it was with no little pride that A Squadron recorded in 1954 that it was SQMS Collins who had put their *Sous-Officier* to bed, emerging from the encounter with a magnificent blue burnous!

There were, in fact, many who welcomed a trip into the 'blue' just for the hell of it. In October 1954, Major Browne and several members of the Regiment accompanied an oil exploration party sent out by the Socony Vacuum Oil Company. Altogether, this covered 5,000 miles of barely mapped territory, expert navigation being provided by a former member of the Long Range Desert Group.

It had long been realised that better training areas existed south of the escarpment in an area known as the Jebel, which was a high plateau running parallel with the coastline and some 50–100 miles from it. The problem was that while a road had been pushed through the crumbling cliff-faces, the tanks would have destroyed it had they used it; further, the tank experts of the Second World War had considered the escarpment itself to be unclimbable. The Regiment did not agree and, knowing the Centurion to be a prodigious hill-climber, A Squadron, commanded by Major Tayleur, had actually found a way up at Garian in November 1954, and returned by the same route to tell the tale. The following January, the entire Regiment performed the climb, led by C Squadron. Mr Tappin, the United States' Ambassador to the Kingdom of Libya, came along to see the fun and actually rode in the leading tank. As a member of the Diplomatic Corps he may not have fully appreciated the full horror of such terms as crabbing and rolling, but the sheer drops beside the track and the painful reversing required at hairpins, with the tanks partly suspended above empty space, told their own story. The climb was completed without casualties, although one tank ran away backwards, being mercifully brought up short by its own mono-trailer, which jack-knifed and buckled beneath it.* Once on the Jebel, the Regiment took part in six weeks continuous exercises with the newly-formed Libyan Army.

One unusual task which the Regiment was required to perform in Libya was evaluation of two prototype Caernarvon heavy tanks in hot weather conditions, and a Trials Troop under Captain R. A. McClure was formed for this purpose in October 1953. The Caernarvon was an experimental vehicle which eventually led to the Conqueror, the largest tank ever built. The work was interesting and rewarding, revealing the traditionally diverse viewpoints of designer and user, concerning which some visitors from the relevant Ministry received a good three-hour heckling from the crews.

In 1955, the Shooting Team achieved a spectacular series of successes in the annual ARA competition, winning the King George V Cup, The Royal Irish Cup, The Young Soldiers' Cup, The Squadron Shield, The First Army Cup, The Duke of Connaught Cup and the coveted Queen Victoria Trophy. It was rare for the last to be won by the cavalry regiment, and the 14th/20th had not done so since 1924. In 1956, the Regiment again won The Queen Victoria Trophy, this time on aggregate, plus the Duke of Connaught Cup and The Royal Irish Cup. This was undoubtedly good for the Army's musketry in general, for the infantry sharpened up their performance to such an extent that when the Regiment next fired in the competition it returned the same high scores which had won two years running yet received only high placings.

The Regiment returned home briefly in October 1955 and moved to Munster the following spring, joining the Twentieth Armoured Brigade. In the summer of 1957, it moved to Hohne near Hanover and became the armoured regiment of the 11th Infantry Brigade. Shortly after it received the unwelcome news that it was, in effect, to be reduced for the time being to an administrative unit only. A and C Squadrons were re-equipped with Saracen APCs and posted respectively to the 20th Armoured Brigade Group at Detmold and to the 7th Armoured Brigade at Celle, while B Squadron, which had much the best of the deal, replaced the Royal Tank Regiment's Independent Squadron in Berlin. These roles were maintained until November 1960, when the 14th/20th assembled again at Hohne.

Few regard this period with satisfaction, for although the APC squadrons performed their task diligently and well, a majority of Hawks believe that it was a waste of resources to employ an operational armoured regiment in this way when other troops could have performed the task with a little instruction. Nor did it help that the Army still tended to regard APCs as a sort of armoured bus service, nor that wheeled APCs, however effective they might be on jungle roads or in an urban counter-insurgency situation, could hardly keep pace with the tanks they were intended to support on an open European landscape.

For much of this period the Regiment was commanded by Lt-Colonel P. F. W. Browne, DSO, MC, who had contributed so much to its history during the Second World War. He is always remembered kindly and as a rugged individualist who tended to make up his own rules. Brotherton would have warmly applauded his thirst for action, and he would certainly have commented that 'he was a delightful fellow, but singular in his dress', as he had done of one of his own Commanding Officers. The fact was that Browne had little use for some of the conventions of peacetime soldiering, and he can be said to have invented Barrack Dress long before the War Office, as it then was, had even thought of it, going

---

*The mono-trailer was a one-wheeled armoured trailer which gave the Centurion additional fuel capacity. The idea was not a success and it was abandoned in favour of an additional armoured fuel tank welded to the vehicle's stern plate.

about his daily round in a well-ventilated brown sweater and a comfortable pair of slacks. Visiting senior officers, whom he regarded as a nuisance, were sometimes told that the Colonel was 'out' unless it was known that they had something to say that was in the Regiment's best interests. Colonel Browne retired from the Army in 1959.

In 1960, the 14th/20th again became an armoured regiment, equipped with Centurions plus one Heavy Troop of three Conquerors in each sabre squadron. The Conqueror mounted a 120mm gun and had been evolved for the fire support role as a direct counter to the Russians' 122mm-armed JSIII and T10 heavy tanks. The latter proved to be rather less of a threat than had once been imagined and soon disappeared altogether from the Soviet order of battle. Nonetheless, the design experience had been extremely useful and contributed much to the Centurion's replacement, the Chieftain.

That same year the escalating cost of armaments forced the government to opt for small, selectively recruited Regular armed services equipped with high-technology weapon systems, and the ending of National Service was announced. This meant that within two years the majority of units would lose a high percentage of their trained personnel and, to counter this, RAC regiments were given specific areas of the country in which to concentrate their future recruiting. The Regiment was particularly fortunate to have been allocated Lancashire, not simply because it had fought its first battle there, nor even because of its long-standing affiliation with the Duke of Lancaster's Own Yeomanry, whom it had supplied with Permanent Staff Instructors since the Territorial Army was reformed after the Second World War, but because the county has always raised more soldiers and regiments per acre than any area of comparable size in the United Kingdom. During the first year of the new policy no less than 200 Lancastrians joined their county's cavalry regiment as regular soldiers.

The old rules governing whether or not a cavalry regiment was eligible to carry a guidon had long since ceased to have any meaning and it was now decided that they should again be issued to light cavalry regiments as being jointly symbolic of the Sovereign herself and the regimental honour. At a parade held at Hohne on 10th June 1961, two old guidons belonging to the Fourteenth and the Twentieth were trooped off the square and Field Marshal Sir Gerald Templar presented the 14th/20th King's Hussars with their first guidon on behalf of Her Majesty The Queen. This had a crimson field with the Prussian Eagle in the centre surrounded by a crowned wreath, the White Horse of Hanover in the first and fourth quarters, and the legends XIV KH and XX H in the second and third quarters respectively.

In 1962, the Regiment converted to armoured cars and embarked for Libya again, to find that recent oil discoveries had induced a degree of prosperity that had been lacking during its last tour of duty there. Of this the most obvious symptom was the disappearance of the stately horse-drawn gharri in favour of the modern motor cab, driven dangerously with obligatory horn noises and a barrage of Arabic expletives. A and B Squadrons, equipped as conventional armoured car squadrons with 76mm-armed Saladins and Ferret scout cars, were based respectively at Benghazi and Tripoli, but were able to exchange stations and training areas for the sake of variety. C Squadron, commanded by Major W. D. Garbutt, was equipped solely with Ferrets and designated air-portable, being based permanently in Benghazi.

Cyprus had been granted its independence in 1959, but this in no way diminished the deep-seated historical hatreds which existed between its Greek and Turkish communities. In December 1963, these flared into open hostility involving loss of life and destruction of property, centred on Nicosia. At 0500 hours on Boxing Day, C Squadron was placed on

Ferret Scout Car and Saracen APC of an Assault Troop, photographed in the desert near Sabratha, Libya.

immediate notice to fly to the troubled island as part of a Truce Force, and was airborne that night. The squadron remained in Cyprus for five weeks, working with the infantry battalions of the Truce Force, or performing independent missions. The situation was handled with a firm impartiality which left both factions in no doubt that a ceasefire would be imposed, by force if necessary; C Squadron came close to opening fire on several occasions but was not required to do so as the opposition clearly understood that it meant business, and did not press matters to a conclusion.

Much of the training around the Benghazi Bulge enabled squadrons to retrace the fighting which had taken place in the area during the Second World War, notably the route taken by Combeforce, the spearhead of the dramatic thrust by the 7th Armoured Division which, in February 1941, had trapped and destroyed Marshal Graziani's Italian army at Beda Fomm. There was still plenty to remind the traveller of the great events which had taken place twenty years earlier; on one occasion an advance supply dump, thought to have been used by Combeforce, was discovered; on another, an uncleared mine blew a wheel off the Saladin which had driven over it. As always, the desert beckoned, and occasionally reminded the unwary that it must never be taken for granted. Major Palmer's A Squadron, during an expedition to Kufra Oasis, discovered an apparently good source of water which was actually a hot spring, capable of inflicting vicious scalds. Without doubt, the Regiment's most enterprising desert traveller was Lieutenant Donald Chappell, who led an expedition to the Uweinat Mountains on the Libyan/Sudanese border in July 1963, and then another across the Sahara to Lagos, Nigeria, in November of that year. The latter covered 6,000 miles in forty-six days, a remarkable achievement bearing in mind that several of the vehicles had been purchased from a dump and virtually rebuilt from scrap.

Adventure training, as this type of exercise is known, is intended to stretch the individual in a variety of ways, and an element of danger is always present. In January 1965, Lieutenant Christopher Cornish, Cpl N. W. Kirkham and Tpr B. L. Bunn were killed by a rockfall while climbing close to the summit of Mount Kenya; the fourth member of the team, L/Cpl J. Crossland, had remained at the highest camp and made the difficult descent of the mountain alone to fetch help when he realised that the others were overdue. The dead were buried with the fullest possible military honours at the foot of the 15,000-feet Darwin Glacier and the Regiment has recently erected a permanent memorial on the spot.

In December 1964, C Squadron flew to Cyprus for a short exercise and the following Easter, A Squadron embarked for a long tour with the United Nations' Peace Keeping Force on the island, during which they were kept extremely busy at Zyyi, Lefka and Paphos. 'We felt ourselves to be doing a realistic and worthwhile job. All had to be relied on to operate as individuals using their common sense in handling some unusual situations. On many occasions there was little opportunity of obtaining guidance from above. This developed everyone's capacity for hard work, their sense of responsibility and their initiative.'

The Regiment returned home at the end of 1965 and concentrated at Perham Down, where it was joined by A Squadron the following March. Because it had been so widely dispersed while abroad, it had been unable to celebrate its 250th anniversary, but this was remedied in May with a series of parties, dinners and balls, as well as a parade. 1966 also witnessed Royal approval for the affiliation of the 1st/6th Queen Elizabeth's Own Gurkha Rifles, who were accorded the privilege of wearing the Prussian Eagle as an arm badge in the same manner already granted to their Second Battalion.

Once more the 14th/20th became an armoured regiment and moved to Paderborn in December 1966, joining the 4th Division. The Conqueror was on the point of being phased out of service, but the Centurion was now armed with the excellent 105 mm gun, used in conjunction with a .50 ranging machine-gun which provided a ballistic match with the HESH round. In 1968, the Regiment converted to Chieftain, which also employed the ranging gun technique in conjunction with its 120 mm main armament. The pace of technical change had now become so fast that more and more was demanded of individual crewmen, and this in itself justified the policy of selective recruiting. There were new and vastly improved radio sets, active and passive night fighting devices and navigational aids, as well as new weapon systems in the form of the Anti-Tank Guided Weapon (ATGW) and the helicopter, the full implications of which had to be thoroughly digested. At this period the Regiment adopted a five squadron organisation incorporating a Command Squadron which included RHQ Troop, the Reconnaissance Troop and an Air Troop. In Libya the Air Troop had consisted of a flight of Auster light aircraft, but in Germany these were exchanged for Sioux helicopters. In 1969 the Air Troops of armoured regiments were formed into Brigade Air Squadrons.

A welcome break from the routine of BAOR was provided by a six-week exercise in Libya during May and June 1969, and it was while the Regiment was there that it received the news that it had been honoured by the appointment of HRH The Princess Anne as Colonel-in-Chief. Such appointments are only made on important Royal occasions, and this coincided with the Investiture of HRH Prince Charles as Prince of Wales at Caernarvon Castle. For the 14th/20th King's Hussars the choice was a particularly happy one, maintaining as it did the tradition begun by the Duchess of York in 1798. The Princess

Armoured car troop of B Squadron returning from exercise through the streets of Hong Kong.

visited her first Regiment in October and promptly proved that she could handle a Chieftain and fire a machine carbine with the best of her soldiers. Since then, she has paid regular visits during which her natural charm, sense of humour, genuine interest and obvious enjoyment have made her many lasting friends in every part of the Regiment.

Towards the end of the year the Regiment again converted to armoured cars and was then dispersed further afield than at any time in its history. A Squadron went to Singapore and B Squadron to Hong Kong, while C Squadron, which again assumed the air-portable role, was based at Tidworth with RHQ.

The troubles in Ulster had now reached new levels of violence and, whatever may have been their original causes, were now being cynically manipulated by the Provisional IRA for their own ends. C Squadron, commanded by Major Peter Harman, embarked for a four-month tour of duty in Belfast in November, and was soon involved in containing some of the most vicious riots the city had ever witnessed, coming under fire several times and being subjected to a hail of missiles, including acid, petrol and nail bombs. The squadron also manned checkpoints, carried out regular patrols and took part in arms searches, capturing a large quantity of ammunition as well as two much-wanted criminals.

A Squadron came home from Singapore in April 1971 and embarked for a tour in Ulster in June of that year, being posted to South Armagh, an area in which the terrorists found a degree of local support. During this period they were based at Gosford Castle, a Victorian

The Regimental Shooting Team which took part in the Army Championships at Bisley in 1972. Left to right, back row: Cpl Farrell, Tpr Smethurst, Tpr Nettleship, Tpr Holden, Sgt Marsh, Tpr Geraghty. Front row: Cpl Ingham, 2Lt Ross, Capt Williams (Team Captain), Lt-Col Williams (CO), Capt Hodson, Sgt Underwood. Trophies include The Unit Championship Shield, RTR Cup, Cambridge Shield, Lindley Cup, RAC Sqn Shield, SMG Cup, KRRC Cup, Pistol Cup.

pile the noble exterior of which belied the decay to be found within. C Squadron and RHQ also arrived in South Armagh in August and it was here that the Regiment sustained its first serious operational casualties. On the 16th, Corporal Peter Webb, who was carrying out a border patrol, was approaching Cullaville Customs Post when a mine, packed with scrap metal, was exploded beside his Ferret. Simultaneously, the car was hit by several bursts of automatic fire which Webb – who later had a ½-inch bolt removed from his jaw and was transfused with nine pints of blood – coolly returned while reversing out of trouble. On 29th August another border patrol, commanded by Sergeant Tottman, ran into a carefully planned ambush near Crossmaglen. The crew of the leading Ferret, Corporal Armstrong and Trooper Agar, were badly wounded – the former fatally – and collapsed in the roadway. Tottman took up the fire fight with the enemy and reported the situation by radio. In due course, a Saracen command vehicle with Corporals McVay and Elsdon aboard arrived at the scene, accompanied by a third Ferret, and Tottman directed the vehicles to form a box around the wounded men. The enemy were present in strength and kept the whole area under continuous fire, which McVay and Elsdon braved to rescue the injured in an operation lasting forty-five minutes. The scale of the ambush was such that the IRA obviously intended to rival the exploits of the Bhoys of Kilmichael, and it was entirely due to the courage and presence of mind of these three NCOs that it failed entirely in its purpose.*

Both squadrons carried out a further four-month tour in Ulster between July and November 1972, A being based in Albert Mill, Belfast, and C at Aldergrove airport. Service in Ulster was regarded as a challenge which sharpened up responses in an active service environment and the tour went well, with one tragic exception.

On the night of 13th July, 1 Troop A Squadron, commanded by Second-Lieutenant Robert Williams-Wynn, was detailed to support an operation mounted by the 1st Prince of Wales' Own Regiment in the Andersonstown district. On approaching the area, the troop came under heavy fire and Williams-Wynn, a popular young officer of great promise, was killed instantly, the fatal round ricochetting inside the Ferret and wounding his driver, Trooper Brian Hansell, in the back. Although in great pain, Hansell reported the incident clearly over the radio and then, on the arrival of reinforcements, freed the turret hatch, exposing himself further to the enemy's fire. He barely mentioned his own injury and insisted on driving to the hospital, where he was found to be so weak that he had to be lifted from his driving seat. In recommending that he be Mentioned in Despatches, Lt-Colonel D. W. Hanson, commanding the infantry, commented that Hansell's 'coolness under fire and his calm and responsible conduct during the incident have received the highest praise from all who saw it. His courage in minimising the effect of his wound was in the highest tradition of the service and his Regiment'.

In April 1973, the Regiment moved to Herford, Lower Saxony, and became the Armoured Reconnaissance Regiment of I Corps, converting to Scorpion light tanks in July. B Squadron rejoined from Hong Kong in August and in October undertook their first tour in Ulster, with SHQ in Gosford Castle and troops at Berry, Newry and Newtown Hamilton, producing good results. They were relieved by A Squadron in January and a most interesting four months ensued, during which the terrorists' grip on Newtown Hamilton was loosened to such an extent that the Royal Ulster Constabulary was able to commence its own patrols in the area again, which it had been unable to do for the previous year. This little town lies only three miles from the border and had been a particular target of the IRA, but the garrison, of which the squadron provided the major part, had restored civilian confidence in the ability of the Security Forces and produced a useful flow of intelligence from the inhabitants. Captain J. N. Symons, Sergeant Redmond and Corporal Tyson were all commended for their constructive contributions to the improved situation, which was in itself symptomatic of the fact that the Security Forces were slowly winning the battle against terrorism.

On 5th March, Second-Lieutenant Colin Wetherell-Pepper's troop, which was manning a checkpoint at Newry on the main Belfast to Dublin road, scored a major success against the terrorists. On receiving a report that several men had been seen behaving suspiciously near the old Customs Post on the border, Wetherell-Pepper set out with two vehicles to investigate. As he approached the Customs Post he saw a coal lorry and a civilian car parked nearby. Three armed men jumped from the lorry and ran away across the fields towards the border, while three more men attempted to drive away in the car. The troop leader gave chase and rammed the car, bringing it to a standstill. The occupants were ordered out, and

---

*During the earlier Troubles a small Security Force patrol was ambushed at Kilmichael and eventually forced to surrender, the affair being complicated by the simultaneous surrender of many of the Bhoys. An incident beloved of Republican balladeers, who are understandably selective in their narrative.

Troop manning a fortified post near Londonderry, 1973.

then told to remove the primed bomb which had been discovered inside, together with a pistol. Meanwhile, the three armed men had taken cover and opened up on the patrol, which promptly returned their fire; of this group one was hit and possibly killed, while the two survivors were apprehended by the Irish authorities and later imprisoned. On conclusion of the engagement the three prisoners already secured were marched back to the lorry and made to remove another primed bomb from the vehicle: they, too, were sentenced to a lengthy term of imprisonment. Wetherell-Pepper received a well-deserved Mention in Despatches for bringing the affair to such a neat and speedy conclusion.*

World interest was now centred elsewhere than on Ulster. In 1974 the animosities between the Greek and Turkish populations of Cyprus had again erupted into violence and Turkey had launched a full-scale invasion, occupying 40% of the island. A precarious cease-fire was maintained along what was known as the Green Line, with the assistance of UN troops and the British Sovereign Base garrisons. In June 1975, A Squadron, commanded by Major J. R. Clifton-Bligh, flew out to join the latter, having converted back to armoured cars, and remained until November. For part of its length the Green Line ran parallel with the boundary of the Eastern Sovereign Base and the squadron was kept busy patrolling this and dealing with refugee problems.

C Squadron, under Major Harman, arrived to join the UN contingent in September and did not return to Herford until March 1976. The squadron was engaged elsewhere along

*Wetherell-Pepper was killed in a traffic accident in Cyprus the following year. He was a keen all-round sportsman who was Regimental Musketry Officer and Athletics Coach.

HRH The Princess Anne, the Regiment's Colonel-in-Chief, at the controls of a Scorpion light tank.

the Green Line controlling an extremely touchy situation in which both sides sought to improve their positions or secure some minor advantage which would confer face, often in rather childish ways. There were times, too, when UN patrols were actually accused of spying for one side or the other. The Turks were marginally the worst offenders, partly because they felt they were arguing from a position of strength, and partly because they believed that the UN troops were naturally prejudiced in favour of the Greeks. This latter assumption was based on the fact that the UN had, with astonishing insensitivity, appointed an Indian officer, Lt-General Prem Chand, to command its force in Cyprus, despite the Indian government's open support of the Makarios regime; in the event, Prem Chand proved to be a firm and scrupulously fair commander. Nonetheless, it was no easy matter to convince the two factions that the word 'No' meant what it said, and that no amount of threats, brandishing of cocked weapons, or even peevish attempts at manual strangulation by unbalanced Turkish subalterns, was going to alter the fact.

The apparently endless stream of defence cuts imposed since the Second World War had by now reduced the strength of the armed services to a level which many officers considered to be dangerously low. The Royal Armoured Corps had suffered as seriously as any

HRH The Princess Anne, escorted by Lt-Colonel J. A. Pharo-Tomlin, talking to D Squadron tank crews during her visit to the Regiment in 1978.

branch of the service but had avoided the loss of two more of its major units by means of a compromise in which regiments manned, by rotation, both the RAC Centre in Dorset and the Corps' basic training establishment at Catterick. In 1976 it was the turn of the 14th/20th King's Hussars to administer the former and in May the Regiment left Herford to assume this role, RHQ, B and HQ Squadrons being based at Bovington, C Squadron at the Gunnery School at Lulworth, and A Squadron with the School of Infantry at Warminster, where it provided demonstrations and training in infantry/tank co-operation.

In October 1977, the Regiment moved to Hohne and again became an armoured regiment, adding a fourth sabre squadron to its establishment, which was gratifying. Rather less gratifying was the loss of its Reconnaissance Troop, the theory being that reconnaissance would be carried out under the centralised control of higher formations. It is, of course, quite contrary to fundamental principles that a cavalry regiment, whether horsed or on tracks, should operate without its own scouts, and this radical decision has recently been reversed, the re-formed Reconnaissance Troop being equipped with Scorpion light tanks.

The situation in Ulster had improved considerably and much of the Province had now been returned to the control of the Royal Ulster Constabulary. Even so, there were areas in which a military presence was still vital and in October 1978 the Regiment flew to Belfast for a four-month tour as an infantry battalion in the city, where it operated in Andersonstown, Lenadoon, Suffolk and Twinbrook. These were predominantly Roman Catholic

Major-General J. M. Palmer,
Colonel of the Regiment since 1981.

Sir John Nott, then Secretary of State for
Defence, receives some sound advice.

areas which had once provided bedrock support for the Provisional IRA and during this period a major effort was made to win the trust and confidence of the population. This succeeded in producing much useful information and in consequence terrorist activity was significantly curbed. Naturally, the dissident elements did not take kindly to this erosion of their infrastructure and two members of the Regiment, Lieutenant Richard Shirreff and Sergeant G. Hutchinson, earned special recognition for their constant pursuit of this objective in particularly dangerous circumstances.

The year 1978 also witnessed the appointment of Major General J. M. Palmer as Director Royal Armoured Corps, the first Hawk to hold this immensely important office and also, incidentally, the first to attain General's rank without service in another regiment.

Lack of sufficient suitable areas for formation exercises in Western Europe meant that battlegroups now trained regularly in Canada. The Regimental battlegroup as a whole first trained there in 1979, followed by B and D Squadrons in 1980, and A and C Squadrons in 1983. On the technical front the need to accelerate target acquisition time and increase the probability of a first round hit to a near certainty had led to the development of the Improved Fire Control System (IFCS), which allowed for such hitherto imponderables as, *inter alia*, trunnion tilt, barrel wear and ambient temperature. In 1980, A Squadron carried out troop trials with this remarkable equipment and quickly demonstrated its accuracy to such effect that a miss became so rare as to provoke serious discussion.

On 12th June 1982, the Colonel-in-Chief presented the Regiment with a new guidon on

The Colonel-in-Chief reviews her Regiment after presenting a new Guidon, Hohne, 12th June 1982.

behalf of Her Majesty The Queen. This differed slightly from the first guidon in that the second quarter was emblazoned with the Red Rose of Lancashire in recognition of the Regiment's home, and the third with Crossed Kukris, symbolising its affiliation with the 6th Gurkha Rifles. The ceremony took place during a memorable parade at Hohne, and since the Hawks tend to do things with some style, it coincided with a spectacular hot-air balloon rally and the issue of a special commemorative postal cover, as well as the usual balls and parties. By tradition, the old guidon becomes the property of the Commanding Officer and on this occasion Colonel Harman presented it to the Sergeants' Mess, where it is now on display.

Pleasant though these events are to recall, the Regiment's collective thoughts were centred on the South Atlantic, where the Falkland Islands were being re-captured in one of the most successful inter-service operations in British military history. This was a campaign in which the 14th/20th King's Hussars, and indeed every regiment in the Royal Armoured Corps, would gladly have given their eye teeth to have participated, but it was not to be. However, subsequent analysis of events suggested that the Argentines had listened to and understood rather more of the British radio traffic than they were meant to and in 1983 the Regiment carried out trials with a new and more secure voice procedure which is to be introduced into general service. This was accompanied by the usual round of individual, troop, squadron and battlegroup training in which good results were obtained by those attending courses at the RAC Centre, and outstanding results during annual gunnery training. As a measure of the Regiment's high morale and standard of discipline it is pleasing to note that on average only one soldier per fortnight appeared before the

Lt-Colonel Peter Harman, Commanding Officer 1982–84.

Commanding Officer, and that of these less than one-quarter of the cases required more serious consideration. During the year, exchanges were made with the 6th Queen Elizabeth's Own Gurkha Rifles, the Royal Canadian Dragoons, the US 11th Armoured Cavalry and HMS *Amazon*.

Adventure training and sport continues to form a major part of the Regiment's programme. Recently, for example, a party visited the High Atlas in Morocco. In 1983, the Regiment won the Tickle Trophy for physical fitness, judged by means of spot checks on every unit in BAOR. It also won the Cavalry Cup Hockey for the second year in succession, came First in the RAC Kiel Regatta, reached the semi-final of the Cavalry Cup Rugby, won a number of events at Hunter Trials, and reached the regional finals of the Inter-Regimental Polo competition.

A professional view of the Regiment today is provided by Brigadier D. M. Naylor, MBE, the commander of 22nd Armoured Brigade. 'During my two years in Hohne it has provided squadrons to support other units in the Brigade in the formation of all-arms battlegroups while, at the same time, forming its own battlegroup. Latterly, the 14th/20th King's Hussars' battlegroup has comprised one company from the 2nd Battalion Coldstream Guards and one company from the 3rd Battalion The Royal Green Jackets, in addition to a squadron from the Regiment itself and a variety of smaller supporting arms. All my dealings with the 14th/20th King's Hussars have combined to give me the impression of a very tight-knit family regiment whose deep roots in the Lancashire area give it great strength. I have always found all ranks to be extremely cheerful and willing and, whatever the task, they tackle it with a gusto not always to be found elsewhere. In the field the Regiment has achieved a high standard of technical and tactical competence and the

Tank commander, 14th/20th King's Hussars, 1984.

results of its Gunnery Camp over the last two years bear this out. I believe the 14th/20th King's Hussars to be a well-motivated and effective regiment and I have felt honoured to have them under my command.'

The object of the study of history in general, and that of military history in particular, is to profit by the lessons of the past in dealing with the problems posed by the present and future, and perhaps it is fitting that the last words in this story should come from an equally detached and professional source, Major John McNulty, an American Armoured Cavalry officer who commanded one of the Regiment's squadrons in 1983/84 during an exchange visit: 'This era of high technology finds the Regiment equipped with laser range-finders, wire-guided missiles and integrated fire control systems and one might conclude that while such things greatly enhance its fighting capability they also tend to depersonalise the role played by tradition in defining its professionalism. That would be a grave mistake. The Regiment has maintained its distinct personality, nurtured on pride, professionalism, consideration for others and, especially, tradition. Every member, both old and new, is acutely aware of the traditions which combine to make up its history, and he is extremely proud of them. Why, then, should this be so? The fact is that the Regiment is its soldiers' home and they will let nothing tarnish its image. Therefore, they try harder to be better – better trained, better prepared and better socially – than other regiments. Their professionalism is a commitment to excellence, a prevailing attitude, in short, a way of life. The men of this Regiment are professional, dedicated, loyal and, above all, fiercely proud of the fact that they are 14th/20th King's Hussars.'

Appendix A: Colonels of The Regiment

1922–1930 Major General Sir Henry W. Hodgson, KCMG, CB, CVO, 14th King's Hussars
1922–1937 General Sir George de S. Barrow, GCB, KCMG, 20th Hussars
1937–1947 Brigadier F. B. Hurndall, MC
1947–1957 General Sir Richard L. McCreery, GCB, KBE, DSO, MC
1957–1966 Colonel R. J. Stephen, MBE
1966–1972 Colonel B. B. N. Woodd
1972–1977 Colonel R. P. D. F. Allen, MBE
1977–1981 Major General P. B. Cavendish, OBE
1981– Major General J. M. Palmer

Appendix B: Commanding Officers of The Regiment

1922–1925 Lt-Colonel J. G. Browne, CMG, DSO, 14th King's Hussars
1925–1929 Lt-Colonel F. B. Hurndall, MC
1929–1932 Lt-Colonel C. G. Darley, DSO
1932–1936 Lt-Colonel J. A. T. Miller, OBE
1936–1940 Lt-Colonel A. V. Pope
1940–1943 Lt-Colonel H. S. Groves
1941–1943 Lt-Colonel J. B. Norton, while actually commanding 26th Hussars
1946–1948 Lt-Colonel H. A. R. Tilney, OBE (had previously temporarily commanded from 1943–1945)
1948–1951 Lt-Colonel R. J. Stephen, MBE (had previously temporarily commanded from 1945–1946)
1951–1954 Lt-Colonel B. B. N. Woodd
1954–1956 Lt-Colonel R. P. D. F. Allen, MBE
1956–1959 Lt-Colonel P. F. W. Browne, DSO, MC
1959–1961 Lt-Colonel E. G. W. T. Walsh
1961–1964 Lt-Colonel G. A. L. C. Talbot
1964–1966 Lt-Colonel S. R. M. Frazer
1966–1969 Lt-Colonel P. B. Cavendish, OBE
1969–1972 Lt-Colonel J. M. Palmer
1972–1974 Lt-Colonel T. G. Williams, MBE
1974–1977 Lt-Colonel W. J. Stockton, OBE
1977–1979 Lt-Colonel J. A. Pharo-Tomlin
1979–1982 Lt-Colonel D. L. de Beaujeu
1982– Lt-Colonel P. Harman

Appendix C: Adjutants of The Regiment

1922–1925 Captain G. M. Hamer, 14th King's Hussars
1925–1927 Captain F. P. Macintyre, OBE
1927–1929 Captain V. H. Jones
1929–1932 Captain D. S. Frazer
1932–1935 Captain J. D. G. Chaytor
1935–1938 Captain B. B. N. Woodd
1938–1940 Captain R. P. D. F. Allen
1940–1941 Captain R. J. W. McAllen
1941 Captain J. E. Parry-Crooke
1941–1944 Captain G. A. L. C. Talbot
1944–1946 Captain B. C. L. Tayleur
1947–1949 Captain A. T. Sturt
1949–1952 Captain W. A. L. Reid, MC
1952–1953 Captain R. A. McClure
1953–1955 Captain J. M. Palmer
1955–1956 Captain P. L. J. Groves
1956–1957 Captain J. D. Gowlett
1957–1959 Captain W. D. Garbutt
1959–1961 Captain C. C. G. Ross
1961–1963 Captain J. A. Pharo-Tomlin
1963–1964 Captain A. C. Park
1964–1966 Captain J. V. Eyre
1966–1967 Captain R. F. Cornish
1967–1968 Captain D. V. F. Chappell
1968–1971 Captain J. R. Smales
1971–1973 Captain C. M. J. O'Brien
1973–1974 Captain J. F. A. Hope
1974–1976 Captain J. M. D. Moger
1976–1978 Captain R. J. Mann
1978–1980 Captain D. J. B. Woodd
1980–1982 Captain J. C. Cameron-Hayes
1982– Captain A. R. D. Shirreff

Appendix D: Quartermasters of The Regiment

1922–1926 Captain J. F. Best, 14th King's Hussars
1926–1941 Major W. J. Smith, MM
1941–1948 Captain P. T. Drew, MBE
1948–1956 Captain R. M. Roberts
1956–1962 Captain B. E. Moore
1962–1964 Captain R. Boulter
1964–1970 Captain T. Vale
1970–1976 Captain E. Sheen
1976–1978 Captain D. A. J. Williams
1978–1979 Major E. Sheen
1979–1981 Major J. J. Escott
1981– Major G. J. Mitchell

Appendix E: Regimental Sergeant Majors of The Regiment

1922–1925 RSM J. Goddard, MC, 14th King's Hussars
1925–1926 RSM W. J. Smith, MM
1926–1931 RSM S. W. C. Piper
1931–1937 RSM L. E. Trueman, MM
1937–1940 RSM P. T. Drew
1940–1941 RSM G. H. Swallow
1941–1942 RSM A. E. Crocker
1942–1945 RSM A. E. Slatter, MBE
1945–1948 RSM C. G. Smith
1948–1952 RSM R. Saville
1952–1953 RSM L. R. Charlton
1953–1954 A/RSM A. E. Le Maitre
1954–1955 RSM B. E. Moore
1955–1961 RSM A. F. Prevett
1961–1962 RSM E. Sheen
1962–1964 RSM P. C. W. Witney
1964–1966 RSM T. R. Cripps
1966–1969 RSM S. Jude
1969–1971 RSM W. J. Bingham
1971–1973 RSM J. Sharp
1973–1975 RSM V. Colbourne
1975–1978 RSM E. Morris
1978–1979 RSM B. Stocker
1979–1981 RSM P. Midgley
1981–1983 RSM B. Draper
1983– RSM J. Schofield

Appendix F: The Regimental Medal

This is awarded to any officer, warrant officer, non-commissioned officer or man of the Regiment who is deemed by the Regimental Medal Committee to have contributed in some conspicuous manner to the military efficiency or the military honour of the Regiment; length of service or attainment of a certain rank does not in itself count as sufficient qualification. The award was instituted in 1909 by Lt-Colonel E. D. Browne-Synge-Hutchinson, VC, thereby reviving a lapsed custom of the 14th Light Dragoons; a separate award, a Bar to the Medal, was instituted in 1946 by Lt-Colonel H. A. R. Tilney, OBE, and is granted in circumstances where the recipient has again rendered such service as to merit the award of the Medal in a new sphere of Regimental life.

Recipients of The Regimental Medal 1909–1983

1. First impression bought from the maker by Lt-Colonel E. D. Browne-Synge-Hutchinson, VC
2. SSM L. Dove
3. Sgt J. Goddard
4. L/Cpl H. Asquith
5. L/Cpl F. Crossley
6. Pte W. G. Bye
7. Saddler Sgt J. Spring
8. Lt-Colonel E. J. Tickell, DSO
9. SSM R. R. C. Carnegie
10. Retained in the Officers' Mess
11. Bandmaster H. Helmsley
12. ORQMS C. W. Andrews
13. SSM F. Battson
14. Lt-Colonel E. J. Bridges, MC
15. Lt-Colonel H. F. Pridgeon, DSO, DCM
16. Captain and Riding Master L. P. Thwaite

17. SSM S. Mayhew
18. Farrier Quartermaster-Sergeant A. E. Cooper
19. SSM S. J. Gardener
20. RQMS R. C. Dutch
21. Lieutenant A. R. Sturt
22. Major V. E. Mocatta, OBE
23. Sgt A. Hallard, DCM
24. Lt-Colonel J. G. Browne, CMG, DSO
25. SSM J. Procter
26. Lt-Colonel F. B. Hurndall, MC
27. SSM L. E. Trueman, MM
28. RSM S. W. C. Piper
29. Major A. V. Pope
30. Captain D. S. Frazer
31. Lt-Colonel C. G. Darley, DSO
32. Major J. D. L. de Wend-Fenton
33. ORQMS D. R. Wilson
34. Major G. M. Hamer
35. Captain R. A. G. Woodhouse
36. Tpr A. E. Everson
37. Captain (QM) W. J. Smith, MM
38. SSM (RI) H. Brown
39. SQMS W. Northam
40. Lt-Colonel J. A. T. Miller, OBE
41. Captain J. B. Pemberton
42. SSM A. E. Crocker
43. RSM P. T. Drew
44. Sgt J. Astley
45. SSM C. Waller
46. Major J. B. Norton
47. RSM G. H. Swallow
48. Bandmaster R. E. G. Grimes
49. Sgt P. Z. Germains
50. Lt-Colonel L. H. S. Groves
51. Lieutenant T. W. Gill, MBE
52. RSM A. E. Slatter
53. RQMS C. G. Smith
54. SSM R. W. Easto
55. Major R. J. Stephen
56. Lt-Colonel H. A. R. Tilney, OBE
57. RQMS R. M. Roberts
58. MQMS C. G. Blacow
59. SSM V. H. Williamson

1st Bar. Major G. H. Swallow

60. Captain R. A. McClure
61. Sgt F. Winstanley
62. Major J. J. Mann

2nd Bar. Captain P. T. Drew, MBE

63. Major P. F. W. Browne, DSO, MC
64. SSM L. J. Adams
65. Major R. P. D. F. Allen
66. Major A. R. Sturt
67. RSM R. Saville
68. TQMS L. R. Charlton
69. Sgt J. W. Walters
70. General Sir Richard L. McCreery, GCB, KBE, DSO, MC
71. Lt-Colonel E. B. Studd
72. Captain P. S. Mosse
73. SSM A. Senior
74. Cpl V. C. Coles
75. Cpl E. A. N. Holdway
76. Lt-Colonel B. B. N. Woodd
77. Lt-Colonel B. C. L. Tayleur
78. Bandmaster R. Hurst

3rd Bar. Captain R. A. McClure

79. SQMS A. E. Cundy
80. RSM A. F. Prevett
81. Sgt J. C. W. Macgregor
82. Sgt G. W. Tasker
83. Sgt D. A. J. Williams
84. Sgt G. P. Shakespeare
85. SSM A. E. le Maitre
86. SSM J. A. Reynolds
87. AQMS J. E. Thompson
88. Sgt W. Volley
89. Major W. D. Garbutt
90. RSM T. Vale
91. S/Sgt J. V. Justin
92. Tpr B. Grubb
93. Lt-Colonel G. A. L. C. Talbot
94. Lieutenant (QM) R. Boulter
95. WO I J. W. Vickers, REME
96. Lt-Colonel E. G. W. T. Walsh
97. RSM E. Sheen
98. Major R. W. English
99. Major C. C. G. Ross
100. RSM P. C. W. Witney
101. Sgt I. R. Rumble, REME

4th Bar. Lt-Colonel G. A. L. C. Talbot

102. Lt-Colonel O. J. F. Fooks

103. Captain G. S. Sanders, ERD
104. Major M. A. Urban-Smith, MC (Rtd)
105. WOII D. J. Flowers
5th Bar. RSM J. C. W. Macgregor
106. SSM W. Taylor
107. Lt-Colonel S. R. M. Frazer
108. Deposited with National Army Museum
109. SQMS C. Osborne
110. Major J. M. Palmer
6th Bar. Major W. D. Garbutt
111. Lt-Colonel P. B. Cavendish, OBE
7th Bar. Major (QM) T. Vale, MBE
112. WOI R. Ottaway, REME
113. SSM J. Sharp
114. Major M. H. Goodhart
8th Bar. Captain D. A. J. Williams
115. Major W. J. Stockton
116. RQMS J. Bury
117. Cpl D. J. Ingham
9th Bar. Colonel B. B. N. Woodd
118. Major J. L. Ferguson, RAPC
119. Captain J. J. Escott
120. Sgt B. Draper
121. Sgt K. Steele
122. Major J. V. Eyre
123. Lt-Colonel T. G. Williams, MBE
124. SSM P. Midgley
125. Sgt C. Tottman
10th Bar. Colonel R. P. D. F. Allen, MBE
126. Major P. Harman
127. RQMS J. Howard
128. Sgt A. Angel
129. S/Sgt J. Diver
130. Cpl A. Smith
131. SQMS J. Schofield
132. Major P. C. Clarke, CVO
133. Captain R. J. Mann
134. Lt-Colonel J. A. Pharo-Tomlin
135. Captain D. J. B. Woodd
136. Cpl D. J. Lowe, ACC
137. Lt-Colonel D. L. de Beaujeu
138. Lt-Colonel J. R. Clifton-Bligh
139. SSM A. G. Wagstaff
140. Lieutenant N. M. Murray
141. Captain J. C. Cameron-Hayes
142. Captain M. H. Burgess
143. RQMS W. M. Butcher

Appendix G: Regimental Affiliations and Alliances

AFFILIATIONS

Royal Navy, HMS *Amazon*
6th Queen Elizabeth's Own Gurkha Rifles
Duke of Lancaster's Own Yeomanry

ALLIED REGIMENTS

Australian Military Forces
2nd/14th Queensland Mounted Infantry
8th/13th Victorian Mounted Rifles

New Zealand Military Forces
Queen Alexandra's Squadron, RNZAC

Zambian Armed Forces
Zambia Armoured Car Regiment

Printed at Picton Print
keyboarding and
page make-up Veronica Pinniger
negatives Mike Simmons
platemaking Yvonne Hallett
machining Neville Fox